AMES 10.00
#4679

Approaches to Paleoecology

ELEMENTS OF ANTHROPOLOGY
A Series of Introductions

Approaches to Paleoecology

Akkaraju Sarma
Temple University

wcb

WM. C. BROWN COMPANY PUBLISHERS
Dubuque, Iowa

ANTHROPOLOGY SERIES

Consulting Editors

Frank Johnston
University of Pennsylvania

Henry Selby
Temple University

Copyright © 1977 by Wm. C. Brown Company Publishers

Library of Congress Card Catalog Number 76-11982

ISBN 0–697–07547-8

All rights reserved. No part of this publication may be reproduced, stored in a retrieval system, or transmitted, in any form or by any means, electronic, mechanical, photocopying, recording, or otherwise, without the prior written permission of the publisher.

Printed in the United States of America

Contents

Preface

Hominid evolution intrigues many people, so much so that nearly everyone at one time or another has wondered about some aspect of hominid evolution. But in spite of this interest, scientists have had little success in focusing on the ecology of prehistoric populations—neither the paleoecology of the earliest hominids nor of later humans first learning how to domesticate plants. There are several reasons for this situation. First, paleoecology as applied to anthropology has developed only quite recently because it depends so extensively on other disciplines. Second, materials for paleoecological interpretations are poorly preserved. And last but not least, the excavators have usually not saved all the materials from which interpretations could be made. Fortunately, all of these shortcomings are rapidly improving.

This book attempts to point out the significance of paleoecology, along with anthropologically significant examples. Also included is a discussion of several source books from which you can gain additional information. All these items have been chosen within the limits of space.

Many persons have contributed to this book, but I wish to single out a few. To my friend and student, Rick Potts of Temple University, I am grateful for so much help. Thanks to Tom O'Connor, who did the drawings. The Department of Anthropology at Temple University extended me several facilities, for which I am thankful. I also acknowledge with thanks the support and assistance given by Professor Francis Johnston and Mr. Robert Nash. I also thank N. Vasantha and Hedy Wassmer for typing the final manuscript. And the work would not have been possible without the understanding cooperation and help of my dear wife, Mrs. Akkaraju Kameswari. To her my thanks.

Akkaraju V. N. Sarma

1 Approaches to the Study of Paleoanthropology: Paleoecological Considerations

INTRODUCTION

Paleoanthropologists seek to understand human forebears and their evolutionary and ecological relationships with other natural phenomena in the ancient past. To do this we must use archaeological evidence to reconstruct hominid ways of life and view them in the ecological context in which hominids lived. By delineating the connections between prehistoric man and his ecological setting, the paleoanthropologist begins to see early man in the context of the total natural world. Human evolution has lately been viewed as a process of hominids adapting both biologically and culturally to changing surroundings. As a result, paleoanthropologists are largely concerned with reconstructing hominid biocultural adaptations to ecological settings.

Describing fossil hominids solely in terms of their physical features deprives us of the original meaning of their ecological adaptations. Paleoanthropologists have recognized the handicaps of these earlier *typological* approaches. Since ecology is emphasized in this book, any environmental reconstruction performed to explain a process will be an ecological reconstruction. Questions posed in ecological reconstructions are processual in nature. For example, when talking of rainfall in an environmental frame of reference we may simply furnish the data as annual amount of rainfall. But in a more processual way we may depict the data as seasonal—in which months the rain falls, how much it affects the cropping cycles, what types of crops are possible, and so on.[1] In an ecological perspective then, we look for adaptive interactions between the sociocultural system and the ecological system, and attempt to understand the relationships involved. Therefore in obtaining data on past or present ecosystems we try to get at points of articulation between sociocultural systems and ecological systems. These systems contain both *variables* and *parameters*.[2] It must also be emphasized that in both past and present ecosystems several possibilities of articulation exist. And as such a particular ecosystem will rarely determine the development of a sociocultural system.[3] In other words, ecology only provides a general framework of possibilities of interrelationships between sociocultural and

1. R. MacAdams, "Archaeological Research Strategies: Past and Present," *Science* 3833 (1968): 1187-92.

2. Alexander Alland, *Evolution and Human Behavior* (New York: Natural History Press, 1967).

3. A. P. Vayda and R. Rappaport, "Ecology: Cultural and Non-Cultural," in *Introduction to Cultural Anthropology*, J. A. Clifton, ed. (Boston: Houghton Mifflin, 1968) pp. 477-97.

ecological systems. Herein, an ecological approach differs from *geographical determinism*[4] or *environmental determinism.*[5] Some of the major problems with environmental determinists' arguments center around their lack of perception of other possibilities. Another implicit assumption of these determinists centers around the idea that the existing ecological or environmental conditions may be applied to prehistoric conditions. They see no significant changes from past to present conditions. However, considerable evidence points to environmental change over the years caused by fluctuations in meteorological and solar phenomena, pedologic, hydrographic, oceanographic changes, and several other variables.[6]

PALEOECOLOGY AND ECOLOGY

The standard definitions of paleoecology are wide and general on one hand, restricted on the other. Paleoecology is the investigation of prior natural events and is based on the principle of *uniformitarianism*—applying present to past conditions and interpreting the application. To do this, it uses many tools from other disciplines. Ager further defines paleoecology as follows: "How and where animals and plants lived in the past is the concern of paleoecology. In other words, paleoecology is the study of the habits of living organisms from the time they first appeared on the earth up until yesterday."[7]

Ecology is "the study of the relationships between organisms and their environment today—of the how's and where's of ephemeral modern life—while paleoecology seeks to answer the same questions for the whole of the living record." Therefore paleoecology is considered by some to be the main subject; ecology, the smaller aspect.[8] Where ecology has those advantages inherent in observing and describing the present state of nature, paleoecology may be characterized chiefly by its "relative lack of control over what can be observed and over the range in space and time that the observables occupy."[9]

Ecology may be broadly considered in terms of *descriptive ecology* and *experimental ecology.* Both ecology and paleoecology form part of the descriptive component. Descriptive ecology:

> seeks objective simplifications of the natural world. It restricts observations to ecosystems in which environmental factors or populations have a suitably small range of variation; it classifies the observations; and it constructs models that simulate the relations among the observations or their classes. In these ways descriptive ecology formulates hypotheses about the interactions of organisms with environment and, by testing these against new sets of observations, develops generalizations of broader application. Experimental ecology, on the other hand, creates its own simplifications of nature by controlling directly as many factors as possible in the situation it observes. The two approaches are complementary, and both are essential.[10]

DIVISIONS OF THE SUBJECT

Since paleoecology is broad and diverse, we should give some order to its contents.

4. Ellsworth Huntington, *Mainsprings of Civilization* (New York: Arno Press, 1972).

5. B. J. Meggers, "Environmental Limitation on the Development of Culture," *American Anthropologist* 56 (1954):801-23.

6. Reid Bryson, "A Perspective on Climatic Change," *Science* 4138 (1974):753-60.

7. Derek Ager, *Paleoecology* (New York: McGraw-Hill Book Co., 1963).

8. Ibid.

9. J. Cushing and H. Wright, *Quaternary Paleoecology* (New Haven: Yale University Press, 1967), p. 2.

10. Ibid., pp. 3-4.

The ecology of individual organisms or small taxonomic contemporary groups has been called *autecology*. When similar work is extended to paleontological specimens, the study is called *paleautecology*. Similarly, studies focusing on contemporary communities as a whole are called *synecology*, while the fossil version is called *paleosynecology*. But in practice, a rigid delineation between individuals and communities is sometimes difficult.[11]

Paleontologists distinguish between evolutionary and ecological approaches. However, in anthropology, since the focus of study is the human species, it would appear that a division into paleoautecology and paleosynecology is superfluous, that the general category of human paleoecology is adequate.

LIMITATIONS OF PALEOECOLOGY

Organic processes of decay as well as erosive geological processes prevent the preservation of much ecological information. Thus paleoanthropologists are limited to investigating only the most ideal areas of preservation. Added problems arise when sediments and other evidence used in ecological reconstruction are moved from the original site of deposition to other regions. When finally unearthed, such evidence may give rise to vague or faulty ecological interpretations since it no longer exists in its original context.

DETERMINATION OF CLIMATIC CHANGE

What is the basis for defining climatic change? On what grounds can the evidence be based and deductions made? Van Rudolff has defined climatic change in terms of observed biological and economic effects in the following translation:

1. A climate change can be spoken of only when . . . the type of weather has so far changed in the area concerned that the makeup of vegetation, the water budget, forms of cultivation and ultimately the fauna are materially changed. . . .
2. After valid climatic change the typical climate of the former period shall only recur in exceptional circumstances (i.e., extreme cases which lie outside the normal variance of the new period). The extreme of [the] old period should no longer recur at all. . . .
3. The new climatic period should stretch over several decades.[12]

When comparing two epochs of the same duration, climatic change may be inferred only if at least two of the above criteria are satisfied. The criteria not only fit recent decades, but are also applicable to prehistoric times. For Van Rudolff's third criterion, the use of absolute and relative dating is extremely helpful in establishing a chronological marker for a period of change. Such criteria as typology, fauna and flora, comparative findings, functional variability of implements, and settlement patterns are helpful archaeologically for establishing time periods into which climatic change data can be tied.

SOME APPROACHES TO THE RECONSTRUCTION OF PAST ECOLOGICAL SYSTEMS

Paleoecology tends to be a pragmatic science. The nature and variety of data needed by paleoecologists to reconstruct past ecosystems require the use of several information-gathering techniques. No single technique has priority; no single one will provide all the necessary ecological information. In anthropological contexts both *artifactual*

11. Derek Ager, *Paleoecology* (New York: McGraw-Hill, 1963).

12. Hans Van Rudolph, "Die Klimapendelungen der Lexten 120 bis 200 Jahre im Sudlichen Oberrheingbiet," *Annalen der Meteorologie* 7 (1956): 12-34.

materials and *nonartifactual* materials are important. Even so, the geographical, geological, and biological contexts are fundamental.

The following is a brief survey of some of the methods used for ecological reconstructions. Artifactual materials are typically exemplified by such objects as stone tools and pottery, i.e., objects that were manipulated in performing past human activities. Nonartifactual materials, though not directly manipulated by humans, played a part in past human activities. For instance, the plant and animal remains associated with human activities are examples of nonartifactual materials. Remains like pollen, shells, bones and pieces of insect exoskeleton, etc. are therefore essential in arriving at an ecological interpretation. They also provide useful clues to food habits, subsistence patterns of prehistoric groups.[13]

Modern environmental data buried in such places as archives, church records, and winery records have ecological value. For example, by using the quantity and quality of grape productions, environmental changes in France's historic period have been documented. In an analogous way large bones, shells, pottery and other larger-sized remains are easily recovered during the screening process in any archeological excavation. Because of their potential value, care should be taken not to reject any materials during the screening process, no matter how trivial the materials may look at first. In the past virtually all of the nonartifactual materials were ignored by the anthropological archaeologist. However, this has changed. All of these materials are systematically collected and analyzed. During soil excavation, samples of it are collected systematically from the section, for both soil and pollen analysis. Some of the methods used in collecting materials for paleoecological reconstruction are provided for guidance. More details can be obtained from the bibliography and further reading recommendations at the end of this chapter.

FLORA AND FAUNA AS SOURCES OF ECOLOGICAL INFORMATION

Pollen grains, microscopic plant materials, offer one of the best sources of ecological information. Pollen analysis involves the study of pollen fossils, which as with other fossils can be identified with existing taxa. Despite the fact that the plant record is so extensive, it is one of the poorest preserved, primarily because of the soft, destructable nature of the materials. The best preserved plant parts are the pollen grains. Since no firm guide exists concerning which parts would be of ecological and cultural significance, a worker should record all the plant parts found—pollen, bark, stems, roots, leaves, flowers, fruits, and fibers at a site.

COLLECTING PLANT MICROFOSSILS

When working at a site, a main aim is to collect uncontaminated samples, taking care not to touch the soil with your fingers. Use a ladle or a broad, smooth forceps. All instruments must be washed after each sample collection to avoid contaminating the next sample. Collection of samples from an exposed vertical profile is preferred. Stretch a tape vertically along the profile and sample at 5 cm-intervals or less. The samples of more complex deposits are best left to a professional palynologist (pollen-and-spore specialist). All samples are collected in sterile containers (like test tubes) and professionally analyzed in a palynology laboratory. Though the subject is complex, it is worth mentioning that pollen production and deposition are governed by such factors as

13. Akkaraju, Sarma, "Holocene Paleoecology of South Coastal Ecuador," *Proceedings of American Philosophical Society* 118, 1, (1974):93-149.

wind direction and rates of pollen production by plants (some are notorious "overproducers"; others, "underproducers"). These factors must be considered in any analysis.

PLANT AND ANIMAL MACROFOSSILS

While plant microfossils must be collected by careful field methods, plant and animal macrofossils are more easily obtained. The most popular technique is called the *flotation technique.* Soil potentially bearing macrofossils is added to a solution of zinc chloride (ZnCl). As the concentration of ZnCl is increased any organic materials present in the soil float to the top where they are skimmed off and later analyzed. By increasing the specific gravity of ZnCl, the materials with lower specific gravities start to move to the surface. Then by using the principle of uniformitarianism, these materials can be identified. In a field situation, where ZnCl is unavailable, common table salt may be substituted. In either case, thoroughly rinse the materials with fresh water to remove the salt content absorbed by the macrofossils.

In general, the plant record is limited because fossil leaves are very rare. Hence specific identifications occur only in rare instances. Pollen grains are most widely used in identifying plant genera. Though the species of seeds and fruits may be identified, plants are ordinarily classified by floral structure. Unfortunately, plant flowers are the most poorly preserved. Analysis of coprolites (fossilized fecal materials) provide valuable information on ecological conditions and subsistence patterns, if chemically rendered into natural consistency.[14] Coprolites found in Mexico have yielded wings and legs of grasshoppers, fibers, shell fragments, pollen grains, and possible squash seeds. Many animals alter their food during digestion, making identification difficult. Carnivores, however, pass hard, almost unchanged parts. Coprolites have been used to reconstruct certain food patterns of the Pleistocene of Australia.[15]

ANIMAL MACROFOSSILS

Generally, all animal remains are macrofossils. As previously mentioned they are recovered during the screening process and through flotation techniques. But mammalian evidence should be used very judiciously in ecological reconstructions. Being warm-blooded, mammals are highly adaptive. However, in conjunction with other supporting evidence, from say, floral information, mammals can provide useful ecological data. The smaller rodents are particularly excellent markers. Another source of valuable ecological information is the dentition of large animals, since teeth closely reflect the environment (food sources and the like) in which the animals lived. In addition they are often the best preserved remains because of their hardness. Many extinct animals are known to us only by their teeth; the famous Canadian anatomist, Davidson Black, was able to establish a primate on the basis of a single tooth! The morphology of crown surfaces has been used in the classification of dentition. With only a few types having been recognized, these are:

brachydont—dentition (when complete set of teeth is present in the animal) is unspecialized, low, and with wide crowns

14. Eric Callen, "Diet as Revealed by Coprolites," in *Science in Archaeology,* D. Brothwell and E. Higgs, eds. (London; Thames and Hudson, 1969): 235-43.

15. Ernest L. Lundelius, "Post Pleistocene Faunal Succession in Western Australia and Its Climatic Interpretation," *Proceedings of the International Geological Congress* 21, no. 4 (1960): 142-53.

typical of a vegetable diet. Example: cattle.

hypsodont—dentition reflects a very highly specialized vegetable feeding habitat. The enamel is deeply folded with all the folds filled with cementum. Example: elephants.

bunodont—dentition displays conical or rounded eminences, less specialized for a varied diet. Example: pigs.

lophodont—dentition shows oblique V- or W-shaped cusps connecting as linear and transverse ridges, for a generalized herbivorous diet. Example: rhinoceroses.

INVERTEBRATE REMAINS

Most invertebrate fossils are mollusks, although the largest animal class among the invertebrates comprises the arthropods. The arthropods are found as bits of fragmentary exoskeletons, which may be compared to modern forms in the framework of uniformitarian principles. Insect fossils may be obtained from archaeological sites by use of ZnCl flotation techniques. But caution must be exercised in selecting suitable samples; many species live in habitats that do not reflect the larger climate of the area (i.e., species of beetles that live in small decaying refuse do not reflect the total surroundings).

Among the insect groups, the beetle group (Coleoptera) is extremely useful for paleoecological information. Other useful finds are small midges (Chironomidae) and grasshoppers (Orthoptera). Some insects can exist only in a narrow range of ecological variation. Such animals are *stenotypic* in character. Thermal factors play an important ecological role in the lives of certain animals. Beetles living in cool temperate conditions, for example, complete a life cycle during summer months, with hibernation during winter. Soil conditions as well as particle sizes are important. For Carabidae beetles the humus content additionally is also important. For many others hygric factors (water availability) are major needs. Some beetles show specific relationships to botanical factors (e.g., the Byrrhidae need moss as food).

Also, since insects as a class possess extreme mobility and rapid reproduction rates, their widespread and rapid dispersals often reflect changing climatic conditions, particularly those associated with thermal changes. The distribution of arthropods is also important. At Trafalgar Square in London, several insects unearthed there are presently identified with southern Europe, particularly dung beetles, Onthofagus, Caccobius, and Oniticellus. This indicates the existence once of warmer conditions at the site.

MOLLUSKS AS ECOLOGICAL MARKERS

Because of their wide distribution mollusks are extremely useful ecological markers. Their usefulness stems from their preservation in a wide variety of environments, their large numbers, and their comparative freedom from human interference. In addition, certain mollusks which are associated with specific environments, and are distributed in specific isotherms, moving north or south due to changing ecologic conditions.[16] Thus they are specific stratigraphic and ecologic markers.[17] In one instance mollusks have been used to reconstruct the paleoecology of the southwestern coasts of Ecuador, specifically in a region where intensive

16. Akkaraju Sarma, "Holocene Paleoecology of South Coastal Ecuador," *Proceedings of the American Philosophical Society,* 118, no. 1 (1974): 93-149.

17. K. W. Butzer, "The Lower Omo Basin: Geology, Fauna and Hominids of Plio-Pleistocene Formations," *Naturwissen-Schaften* 58 (1971): 7-16; ———, *Environment and Archaeology* (Chicago: Aldine Publishing Co., (1969).

archaeological survey was undertaken. Details of this reconstruction on the Santa Elena Peninsula are furnished in chapter 5.

Nonmarine mollusks occur inland—along alluvial deposits, loess (wind-borne glacial deposits), and sediments along the lakes and marshes. The literature is extensive on nonmarine mollusks but much of the data is in the form of faunal lists.[18] Among the nonmarine mollusks, some forms are habitat-specific. The following are examples from western Europe only. However, you could obtain similar evidence in any other region from field observations.

Loess and cold deposits—*Pupilla muscorum, P. columella.*
Marsh forms—*Succinea arenaria, S. oblonga.*
Warmer and interglacial deposits—*Corbicula fluminalis, Helicella crayfordensis.*

In general, neither marine nor nonmarine mollusks migrate unless the ecological conditions are appropriate. Hence a time lag exists between immigration and emigration. The forms will linger on for a brief while, attempting to survive in the changed ecology, but eventually will die out under these unfavorable circumstances. A cautionary note is needed here: account for any local (micro) climatic conditions where they are exceptions to the surrounding general pattern.

Among the marine invertebrates, microscopic forms called *foraminiferans* are particularly useful. They are ocean-dwelling planktonic forms (i.e., they have no independent motion of their own), sensitive to changes in water temperature and salinity. One of the more important forms is *Globorotalia menardii* (four varieties occur), which responds to temperature by coiling clockwise or counterclockwise. By using a mass spectrometer, one can obtain the ratios of the oxygen isotopes, ^{18}O and ^{16}O, to each other, and thereby determine the past ocean temperature.[19] These two sources of information are useful in paleoecology, particularly so when the information can be related to paleoanthropological data.

GEOCHEMISTRY AND UNFAMILIAR GEOLOGICAL PROCESSES

Geochemistry refers to the chemical aspects of geological events. The patterns, mobilization, and cycling of chemical elements and compounds during the earth's history provide clues to ecological conditions. In a given chemical environment such factors as oxidation potential and acidity/alkalinity quantitatively specify the amount of proton activity in elements and compounds. Since all environments are dynamic (subject to changes of climatic conditions) and evolving, the absolute time when certain events occurred is an important factor in understanding these geochemical relationships.[20] Hence by using geochemical relationships, trace elements (rare elements that occur as parts per million, ppm) such as manganese, strontium, and deteurium have been studied as a source of paleoecological information. For example, at Rampart Cave, it was observed that high quantities of manganese (about 1000 ppm) were found in sloth dung deposited during pluvial periods while low amounts (about 200 ppm) was found in deposits of the interstadial peri-

18. B. W. Sparks, "Non-Marine Mollusks and Archaeology," in *Science in Archaeology*, D. Brothwell and E. Higgs, eds. (London: Thames and Hudson, 1969) 313-25.

19. R. F. Flint, *Glacial and Quaternary Geology*, (New York: J. Wiley & Sons, 1971).

20. P. Damon et al., "The Present Status and Future Possibilities of Geochemistry as Applied to Paleoecological Research" in *Reconstruction of Past Environments*. Assembled by J. Hester and J. Schoenwetter, Fort Burgwin Research Center (1964), 77-84.

ods.[21] Such studies have been extended to Niah Cave in southeast Asia. The use of trace-element analysis for archaeological work is only in a beginning stage. Sample collection involves obtaining about 10 grams of soil from uncontaminated, stratified soil.

Studies of trace elements in humans are becoming more common. At least fourteen trace elements have been identified as essential to human health.[22] More speculatively, the availability or limitations of trace elements may have played a role in the domestication of the ungulates; significant behavioral alterations in the animals may have taken place if the availability of trace elements was altered through tethering or coraling.

Geologic processes, particularly volcanic action, may provide information on human ecology in the past. Even in modern days, the devastation produced by volcanic action is noteworthy. In 1974 Hekla in Iceland erupted and produced badland topography. The 1912 eruption of Mt. Katmai in Alaska produced ash deposits, followed by acid rains. People were burned by the rains, vegetation damaged, and metal corroded; erosion and large mudflows followed. Similar phenomena followed the eruption of Paracutin in Mexico in the early 1900s. In the ruins of Pompeii the lava flow had been heavy enough to produce lava casts of all living forms, thus leaving a vivid picture of the extreme devastation that took place. After such volcanic action, the vegetational recovery is very slow and, if the human economy is based on hunting and gathering, the effects of volcanic actions may be disasterous. Even volcanic fumes may be fatal to plants. The fatal limits of some of the fumes are 10 ppm of NH_3 or HCl gases; 1 ppm of SO_2 or chlorine; about 0.001 to 0.01 of Fe. Damage is greater for vegetation when the atmospheric temperature is more than 5°C. Furthermore, high relative humidity accelerates destruction. As mentioned earlier it produces an eroded, or badland, topography. Another source of unusual ecologic information include climate-influenced evaporites, which occur at lake areas. The evaporites from lake areas include carbonates that form sulfates, chlorides, and other salts. The plant communities around a lake basin show the effects of the increasing salinity; survivors could tolerate the increasing quantity of salts. The location of archaeological sites therefore should be viewed not only in the usual geologic contexts but also with an eye toward unusual geologic processes.

HYDROLOGY AND METEOROLOGY

While palynology can provide information on wetness/dryness (through study of the relative numbers of plant materials in a site), and geologic evidence can furnish data on deposition and erosion, not much is known about hydrologic processes that affect the ecology of an area at any point in time. Of considerable importance, however, hydrology and meteorology provide clues to the relationship of life and water, evapotranspiration rates, stabilization of soil surfaces, and regulation of water movement through the biotic communities, as well as provide useful clues to ecological conditions.

In the case of rivers, the forces that produce the flow of water also exert a shear (i.e., a tangential force). This causes the load of the river to move along the shear side of the channel, significantly affecting the river geometry. In a region where rock

21. P. S. Martin, B. Sabels, and R. Shutler, Jr., "Rampart Cave Coprolites and Ecology of the Shasta Ground Sloth," *American Journal of Science* 259, 2 (1961): 102-27.

22. Thomas H. Maugh, "Trace Elements: A Growing Appreciation of Their Effects on Man," *Science* 183 (1973): 254-54, Henry A. Schroeder, *The Trace Elements and Man* (Conn: Devin-Adair), 1973.

formations are fixed, no significant changes take place and the annual runoff (discharge) from the discharge area is a function of the rainfall temperature and vegetation cover. Fluctuations in closed lake levels are related to climatic changes. A lake level fluctuates when the input and output are significantly altered.[23]

Using uniformitarian principles, it is possible to reconstruct paleoecology of an area. In an example cited earlier (and discussed in chapter 5), the marine mollusks such as *Anadara tuberculosa* are habitat-specific to mangroves. Mangroves need over 400 inches of annual rainfall to survive, but at the Santa Elena Peninsula of Ecuador there may be no recorded annual rainfall for several years. For mangroves to have existed in such an area, the cold, desert-producing Humboldt Current, which now skirts the peninsula before swinging out to sea probably once flowed farther south. The equatorial warm waters, including the El Niño, were displaced southwards, thus making it possible for mangroves to survive.[24] Such displacements of currents did take place during 1971 and 1972, making the peninsula a carpet of green, in an otherwise monotonous relief (see chapter 5 for details).

GEOLOGY AND STRATIGRAPHY

The geology of an area and the stratigraphy of a site from a geological point of view furnish information on the age of materials and the nature of depositional history.[25] During the Quaternary period, as during earlier geological periods, the time perspective is important. However, when we come to the historic period, time becomes an "absolute" dimension. That is, in historic times we come to know how old something is in "years." But in earlier periods, time is still a "relative" dimension because the time spans are so great. During both periods, geological strata are deposited on top of each other, with the earliest materials at the bottom followed by later materials at the top. Floral, faunal, and other sources of ecological information are derived from the sequences of these strata (when suitable techniques are used). But to the paleoanthropologist, the Quaternary is the most important period, even though it constitutes only 0.5% of the total time since life first evolved on earth. The Ice Ages constitute one of the major features of the Quaternary. Because of a worldwide rise and drop in temperature, the glaciers expanded and, since much of the Earth's water was locked into the glaciers, sea levels subsequently lowered. The biota associated with these climatic changes offer some of the best indications of changing ecological conditions. It may be worth mentioning that there were two other periods of glacial activity (during the Cambrian and Carboniferous geological periods). Since human life had not yet evolved on the Earth during those times, they are unimportant for our purposes.

The geography of an area provides data on the physical features of the site(s) being investigated. The features of interest include: (1) the site's location relative to "terraces" of the river (sometimes a clue to ages of the samples and ecology); (2) its location relative to old beach lines (information on past fluctuations of sea level and ecology based on associated mollusks); and (3) its location relative to inland lake bodies (information on the fluctuation of lake levels and on the forms of life that lived in the

23. W. B. Langbein, "Hydrologic Tools in Paleoecological Reconstructions" in *Reconstruction of Past Environments.* Assembled by J. Hester and J. Schoenwetter (Fort Burgwin Research Center, 1964): 37-39.

24. Sarma, "Holocene Paleoecology."

25. Flint, *Glacial and Quarternary Geology.*

lakes, especially revealing through pollen analysis).[26]

CULTURE AND ECOLOGY

The distribution of cultural materials at a site can provide excellent information on cultural behavior of inhabitants of particular environments. For example, the earliest australopithecines, living on swamp lake banks in east Africa, transported large boulders (called "manuports") to the areas where they lived. Manuports were possibly used to weigh down materials used as windbreaks. Since the boulders are not local to the region, presence of such large transported stones reflects organized behavior. Therefore in addition to previously mentioned sources of ecological information, culture itself provides useful ecological clues. For example, the abrupt discontinuity of pottery in an area and its reintroduction at a later time shows clearly that abandonment and reoccupation took place—attributable to ecologic changes. In another example, extended residence at a place by hunters, especially by the early hominids, results in a degree of soil alteration. The formation of distinct paleosoils hence may suggest some kind of home base.

While this survey is not exhaustive, it provides some outlines as to the nature of materials available for reconstructing the paleoecology of an area. In particular, remember that ecological reconstruction, as a step in reconstructing the lifeways of a people, requires the information available from a wide variety of sources. Works cited in the bibliography provide more details on the techniques described here.

For Further Reading

Brothwell, Don and Eric Higgs. *Science in Archeology,* 2d ed. New York: Frederick A. Praeger, 1969. Several scholars in different fields have contributed key statements about their subjects in this book. The volume contains papers on palynology, dendrochronology, and other associated fields.

Clarke, David L. *Analytical Archaeology*. London: Methuen and Co., 1968. A book that uses the systems-theory approach to archaeology. Although perhaps difficult to read and comprehend, the introductory student might profit from the case studies in the back of the book.

Coope, G. R. and John A. Brophy. "Late Glacial and Environmental Changes Indicated by Coleopteran Succession from North Wales." *Boreas* (1972):97-142; and,

Coope, G. R. "The Value of Quaternary Insect Faunas in the Interpretation of Ancient Ecology and Climate." *Congress of the International Quaternary Association* 7(1969):359-80. Two important papers about a little-recognized aspect of paleoecology—the use of insects for climatic reconstruction.

Hole, Frank and Robert F. Heizer. *An Introduction to Prehistoric Archaeology,* 3d ed. New York: Holt, Reinhart, and Winston, 1973. This valuable book contains an extensive bibliography.

Kummel, B. and D. Raup, eds. *Handbook of Paleontological Techniques*. San Francisco: W. H. Freeman Co., 1965. A major book containing papers written by several experts on field techniques. Anyone planning to do paleoecological fieldwork should look through this book.

Reyment, R. A. *Introduction to Quantitative Paleoecology*. Amsterdam: Elsevier Publishing Co., 1971. Technical in nature, this recent book uses statistical tools to draw its inferences, while employing experimental data.

26. K. W. Butzer, "The Lower Omo Basin: Geology, Fauna and Hominids of Plio-Pleistocene Formations. *Naturwissenschaften* 658 (1971): 7-16.

Bibliography

Butzer, Karl W. 1969. *Environment and Archeology:An Introduction to Pleistocene Geography.* Chicago: Aldine Publishing Co.

Dimbleby, Geoffrey W. 1967. *Plants and Archeology.* New York: Humanities Press.

Gecker, R. F. 1965. *Introduction to Paleoecology.* Translated and edited by M. K. Elias and R. C. Moore. New York: American Elsevier Co.

Imbrie, J. and Newell, N. 1964. *Approaches to Paleoecology.* New York: John C. Wiley & Sons.

2 Dating Fossil Materials

To understand the hominid evolutionary process we must of course know the age of fossil and archaeological materials. Though spatial relationships—that is, how these fossil materials are distributed geographically—are similarly important, our discussion will center on time relationships.

This section provides a guide to some of the important techniques used in dating pertinent materials. The best source for dating techniques for anthropologists is the work of Michael and Ralph.[1] The types of dating techniques discussed in this chapter are central to paleoanthropological considerations. They involve both absolute and relative methods. When we talk of absolute chronology, we refer to the ages of materials in terms of so many years before the present (B.P.). In relative dating, we may only say which materials preceded or succeeded each other on the basis of stratification, or *seriation.* In recent years, thanks to significant work done in the physical sciences, our understanding of absolute dating has improved tremendously. Some of the dating techniques are discussed below.

PHYSICAL METHODS

Most of these are based on isotopic techniques; the instability of the atomic structure of an element underlies these techniques. The potassium-argon (K-Ar) dating technique is an often-used isotopic technique.[2] It is particularly useful in dating the earliest of the primate materials; much of the chronological information on the earliest hominids from Africa derives from K-Ar dating. For example, there was past volcanic activity at Olduvai Gorge, in eastern Africa where some of the earliest known hominids lived. From the tuffs of volcanic ash that covered the fossil and lithic (stone tool) remains, K-A dates have been obtained. In freshly erupted volcanic materials the amount of radioactive potassium (potassium-40) is relatively high. This potassium-40 progressively decays into argon (argon-40). At any point in time, we may estimate the age of the materials by measuring the ratio of potassium-40 to argon-40. The *half-life* of potassium-40 (the amount of elapsed time since potassium-40 has reduced to half its original amount) is 1.3 billion years. This technique is therefore particularly suitable for materials that are

1. H. Michael and E. Ralph, *Dating Methods for Archaeologists* (Cambridge, Mass.: MIT Press, 1973).

2. J. A. Miller, "Dating Pliocene and Pleistocene Strata Using Potassium-40-Argon-39 Methods," in *Calibration of Hominoid Evolution,* W. W. Bishop and J. A. Miller, eds. (Edinburgh: Scottish Academic Press, 1972), pp. 63-76.

several millions of years old. The use of the K-Ar technique requires special preconditions, further discussed by Hole and Heizer.[3]

In recent years the K-Ar method has yielded dates for important hominid fossil areas. While the dates from Olduvai Gorge will be presented later, some dates from hominid fossil locations in the Omo valley of Ethiopia may be given here:

Formation	Date (B.P.) in millions of years
Mursi	4.05 +/— 0.20
Nkalabong	3.95 +/— 0.11
Usno	2.64 +/— 0.92
	3.31 +/— 0.42

Another isotopic technique of great significance for time determination is the radiocarbon (carbon-14) technique.[4] This method utilizes the low level of radioactivity of carbon found in ancient organic materials. All living objects contain carbon-14—the radioactive isotope of the carbon atom—which after the death of the organism decays to C-12. The half-life of carbon-14 is 5730 years (figure 2.1). With this technique one could measure, under suitable conditions, time depths to about 50,000 years.

Still another dating technique helpful not only in interpreting dates obtained by the radiocarbon method but also in dating old pueblo sites in the American Southwest is *dendrochronology*.[5] Here, the well-known phenomenon of annual growth rings in plants is used. After a *master chart* for a region is prepared (figure 2.2), it can be used for dating the sites by comparing growth rings of the nearby plants. These growth rings are extremely sensitive to climatic fluctuations. During lean years the individual rings are thin; during good years, thick. Long-lived plants like bristlecone pine and ponderosa pine are extremely useful plants for preparing master charts. Dendrochronology is useful for material dating back to about 5000 years B.P. From growth rings it has been possible to check the accuracy of C-14 dates, since the tree rings provide an idea as to the extent of fluctuations of carbon-14 in the atmosphere. Using this principle several scholars have introduced correction factors for radiocarbon dating.[6]

Another technique from which useful dates have been obtained for paleoanthropology is *fission track dating*.[7] This technique is based on the following known facts. Minerals and glasses containing uranium-238 undergo spontaneous fission (the explosive fragmentation of nuclei), which creates a single, narrow, and (by chemical means) microscopically detectable track of a certain density. The fissioning is directly related to the amount of uranium present. Thus experimentally inducing the sample to fission by bombarding it with a measured dose of neutrons and comparing the induced track density with the original track density, the age of material may be calculated. This technique is applicable to material only decades old to some samples millions of years old. It has been used in Olduvai Gorge in Africa to date some of the earliest hominids.

Another useful but infrequently used technique relies on *paleomagnetic reversals*

3. G. Hole and R. F. Heizer, *An Introduction to Prehistoric Archaeology* (New York: Holt, Reinhart, and Winston, 1967).

4. J. R. Arnold and W. F. Libby, "Age Determinations by Radiocarbon Content: Checks with Samples of Known Age," *Science* 110 (1949):678-80.

5. B. Bannister, *Dendrochronology in Science and Archaeology*, rev. ed., D. Brothwell and E. Higgs, eds. (London: Thames and Hudson, 1969), pp. 191-205.

6. P. Damon et al., "Dendrochronologic Calibration of Radiocarbon Time Scale," *American Antiquity* 39, no. 2 (1974):350-66.

7. R. L. Fleischer and H. R. Hart, Jr., "Fission Track Dating: Techniques and Problems," in *Calibration of Hominoid Evolution*, W. W. Bishop and J. A. Miller, eds. (Edinburgh: Scottish Academic Press, 1972), pp. 135-70.

in rocks. In the geological past there were four major reversals in the Earth's magnetic field. Rocks from different strata thus show differences in magnetic polarity. By measuring the magnetic polarity of rocks from different geological strata and dating them

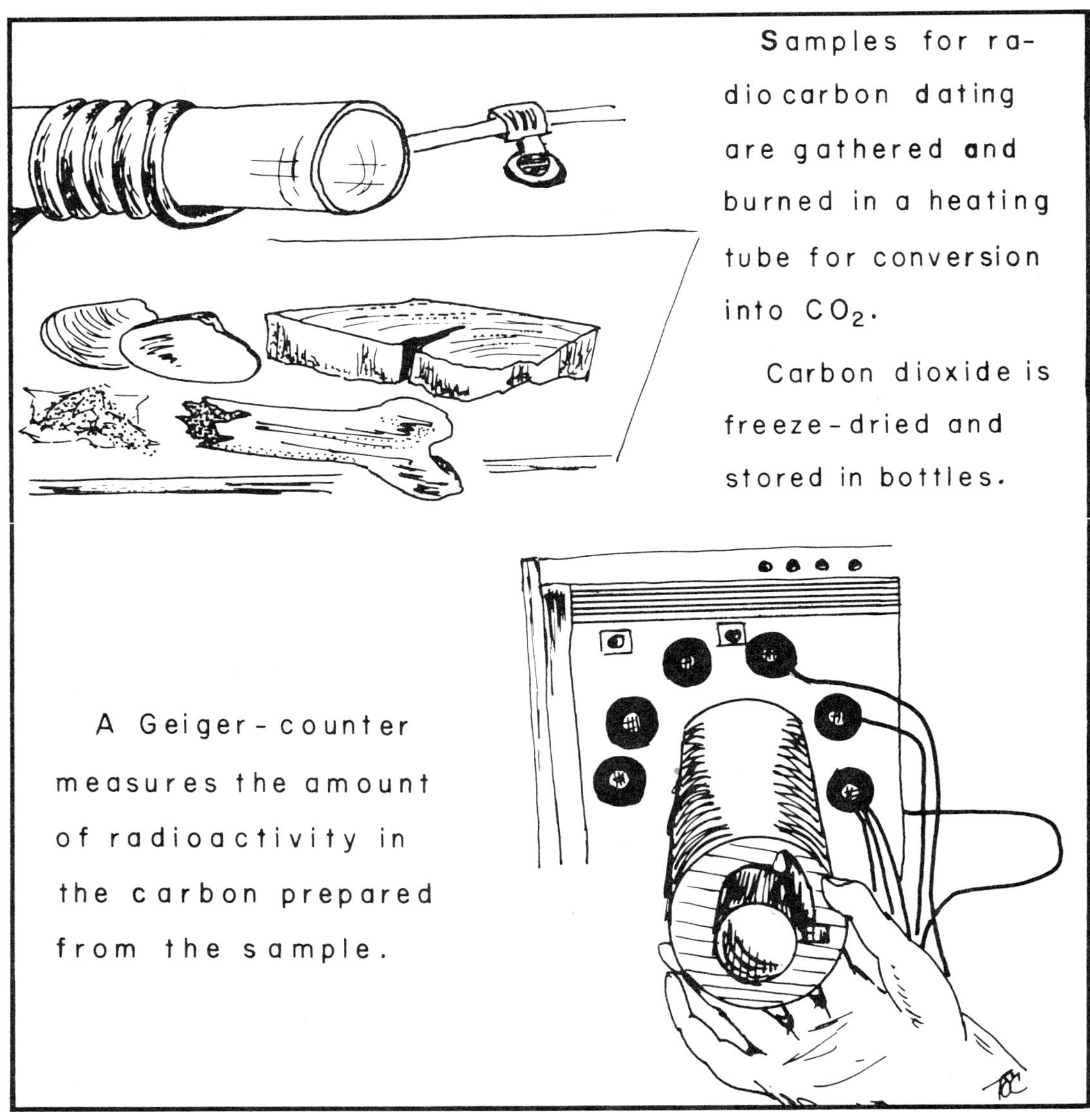

FIGURE 2.1 Carbon 14 as a dating tool. This is a technique that is very useful for dating organic materials from archaeological sites. In the last two decades many improvements and correction factors have been introduced. This technique needs sophisticated equipment.

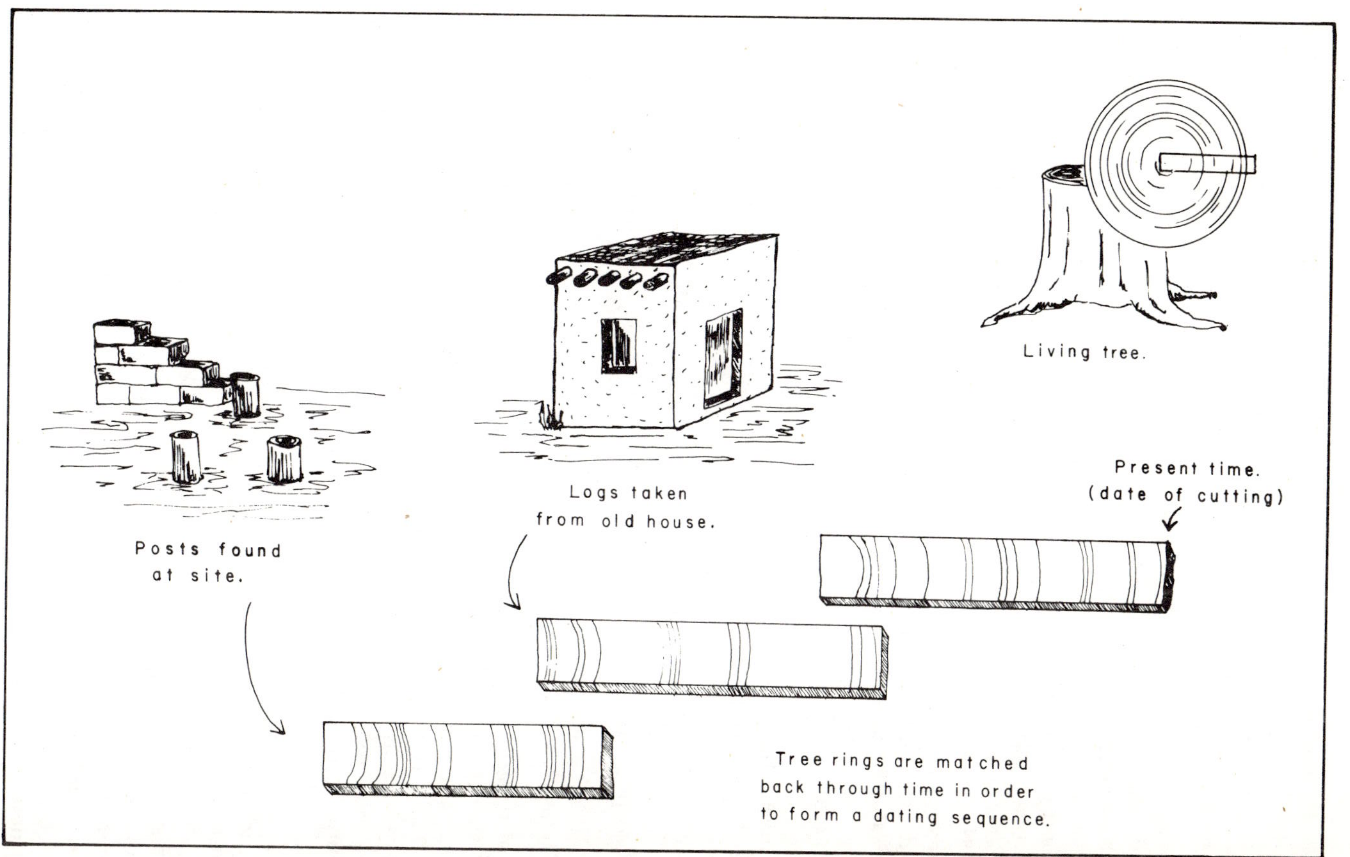

FIGURE 2.2 Dendrochronology as a dating tool. Here the annual growth rings are utilized for dating purposes. A master chart is built for a region on the basis of long records existing in long-lived plants. Once you construct a master chart out of the materials excavated from a site, suitable comparisons, when applicable, can provide information on age. In addition, growth rings are useful as climatic markers; in good years, the annual rings are well developed and for poor years the rings are poorly developed.

by, say, K-Ar technique, certain geological events can be dated.[8]

CHEMICAL METHODS

Recent advances in organic chemistry have suggested the possibility of using *racemization,* the change of optically active chemical compounds to optically inactive ones, as a method of establishing age. This technique has been applied to 100,000-year-old materials. Besides simply dating the materials, this method (using allo-isoleusine/isoleucine ratios) can help in establishing whether paleoanthropological materials were deposited at the same or at different times.[9] However, at temperatures of over 20°C this technique has only limited applicability.

Another technique that has played a role in determining the contemporaneity of fossils employs the absorption of fluorine into bone. The famous "Piltdown man," a fraudulent specimen, was exposed, through use of the fluorine absorption technique.

RELATIVE DATING TECHNIQUES

Techniques of relative dating help determine order of succession, without measuring sample ages in terms of absolute years. For example, in any archaeological deposit, the earliest materials are generally at the bottom and the most recent ones are at the top. This technique (figure 2.3) is based on *stratification,* and the principle extends to geological contexts, because geological strata also reflect their depositional sequence in normal circumstances. Thus a relative sequence through time may be established.

Still another source of relative dating is in the evolution of styles, which change regularly through time.[10] A modern example is the evolution of automobile streamlining (figure 2.4). Not too long ago, most American cars were aerodynamically unsuitable, but since the 1950s this trend has changed. Also, the rear fenders of U.S. automobiles were "finless" in the early 1950s, "developed" fins in the late 1950s through early 1960s, and began to lose their fins in the late 1960s. In all these aspects one could trace a "continuity" in style evolution. Hence it is reasonable to assume that two styles that are close together in appearance are more likely to be contemporaneous than two that are dissimilar. This technique is called *seriation* and is useful for dating archaeological materials, through observation of details of manufacture, shape, and decoration.

THE QUATERNARY PERIOD AND PALEOANTHROPOLOGY

The Quaternary period is sometimes referred to as the *Pleistocene,* the *Ice Age,* or as the *Age of Man.* It is the last major division of geological time following the preceding epoch, the *Pliocene.* The Quaternary period is a time-stratigraphic term. The animals of the Pliocene and their evolution into, and replacement by, modern forms has become important in dating paleoanthropological materials. This replacement of archaic fauna by modern animals (like the modern forms of elephants, camels, horses, and so on) signals the beginning of the Pleistocene time period. The "modern" faunal complex is called the *Villafranchian* fauna, a name reflecting that the original paleontologic observations were made at the

8. A. Cox, "Geomagnetic Reversals—Their Frequency, Their Origin and Some Problems of Correlation," in *Calibration of Hominoid Evolution,* W. W. Bishop and J. A. Miller, eds. (Edinburgh: Scottish Academic Press, 1972), pp. 93-106.

9. K. K. Turekian and J. L. Bada, "The Dating of Fossil Bones," in *Calibration of Hominoid Evolution,* W. W. Bishop and J. A. Miller, eds. (Edinburgh: Scottish Academic Press, 1972), pp. 171-86.

10. J. Rowe, "Stratigraphy and Seriation," *American Antiquity* 26 (1961):324-30.

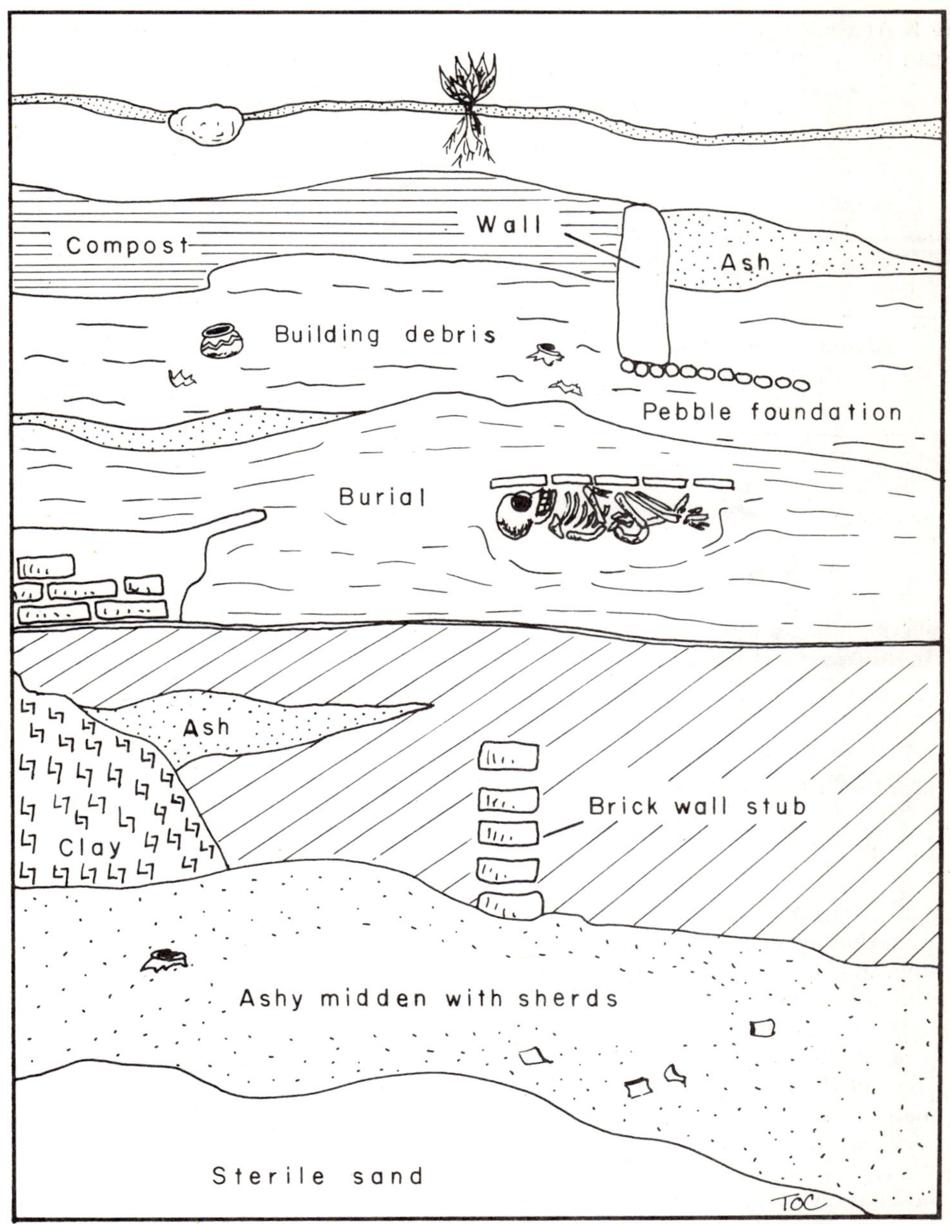

FIGURE 2.3 Stratigraphy as a dating tool. This technique uses the law of super position of strata, and, as such, can only furnish us with relative depositional sequence. The top-most layer being the latest, while the lowest one being the earliest.

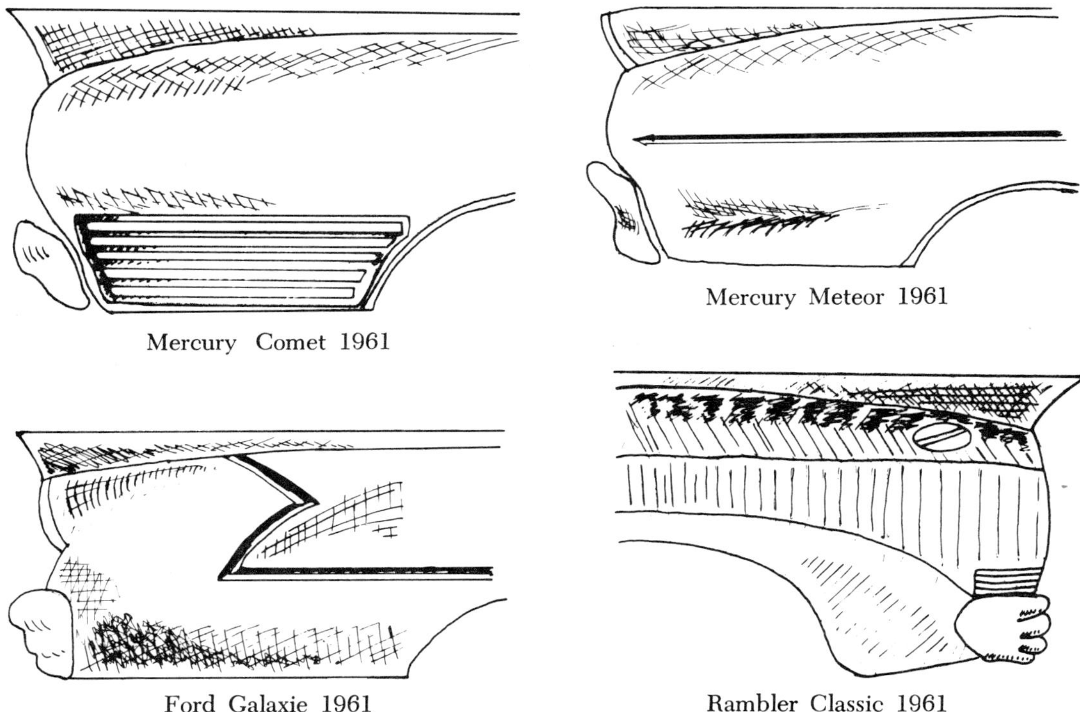

FIGURE 2.4 Seriation of styles. This technique can give useful clues on contemporaneity of styles and relative successions. In the sample shown, one could see similarity of styles (rear fenders of autos) at a time period. The evolution of this styling has not been abrupt. The styles evolve through time. Such evolution of styles in anthropology is very useful. Traits so used are applicable for analysis relating to pottery.

type site of Villafranche-sur-Mer, in France. Not only do the fauna give a clue to the age of the deposits, but also provide information on the paleoecology of the period (figure 2.5). In some areas, where the physical or chemical methods are inapplicable, the only sources of dating available may be the faunal materials, geological strata, and the Quaternary geomorphological evidences. Two geologists, Brückner and Penck, working in the Swiss Alps noted that the glaciers had advanced and retreated four times in the past, leaving behind moraines, high terraces, and transported materials during such advances. On the basis of the elevations of these characteristics, the glacial advances were named Günz, Mindel, Riss, and Würm, from the earliest to the most recent in the sequence. From more recent investigations, it has become clear that the Quaternary and the Pleistocene are not equivalent to one another. Instead, they are independent of each other, with the Pleistocene forming only a part of the Quaternary. It is also clear now that hominid evolution commenced far before the beginning of the Pleistocene epoch. Figure 2.6 points out the main highlights of the Pleistocene time period.

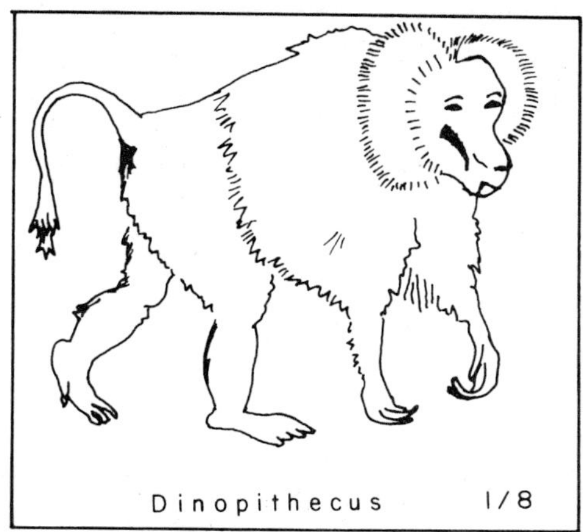

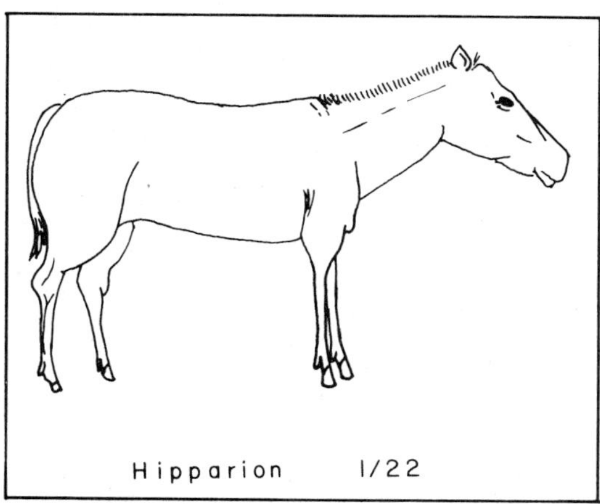

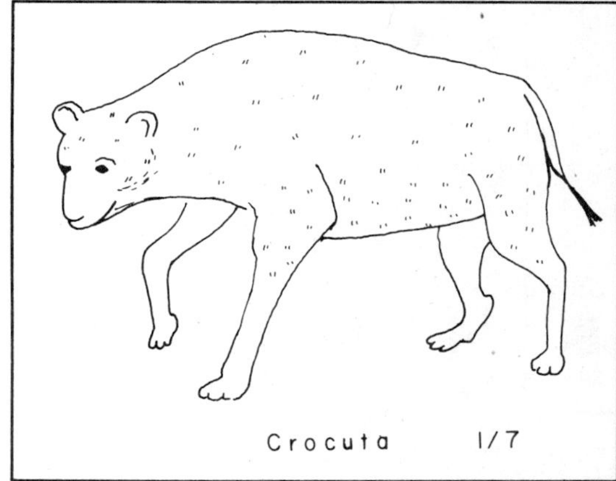

FIGURE 2.5 Villafranchian Fauna. The succession of animal life provides important chronological clues. During the Pleistocene, animals resembling the modern ones came to exist for the first time. These animals (like modern horses, bears, elephants, camels, and others) besides being time markers also furnish ecological clues. The remains of Pleistocene time period mammals have been particularly useful. Most of the animal remains are found at sites in form of teeth, bone fragments, and in some rare instances preserved intact at cold latitudes. The figures in this instance are all reduced to the scale indicated for each.

For Further Reading

Michael, H. and H. Ralph. *Dating Methods for Archaeologists.* Cambridge, Mass.: MIT Press, 1973. An excellent book containing a section on field collection methods.

Bibliography

Ralph, E. K.; Michael, H.; and Han, M. C. 1973. "Radiocarbon Dates and Reality." *MASCA Newsletter*. Philadelphia: University of Pennsylvania Museum, Applied Science Center for Archaeology.

GEOLOGICAL DIVISION	GEOLOGICAL EVENT	APPROXIMATE AGE	HOMINIDS	FAUNA	FLORA
UPPER PLEISTOCENE	Fourth (Wurm) Glaciation	About 70,000 yrs. Ago	Modern Man Neanderthaloids	Modern Animals—Recent Types	
	Third (Eemian) Interglacial	90-000-100,000 yrs. Ago	Neanderthaloids (?)		
MIDDLE PLEISTOCENE	Riss Glacial Complex	At Least 200,000 yrs. Ago	Neanderthaloids (?) H. Erectus		
	Second (Mindel or Holstein) Interglacial	At Least 500,000 yrs. Ago	H. Erectus		
	Second (Elster or Mindel) Glaciation	750,000 To 1 Million yrs. Ago	H. Erectus		
LOWER PLEISTOCENE	First Interglacial First (Gunz) Glaciation	2-3 Million yrs. Ago	Australopithecines	Villafranchian Fauna	
PLIOCENE		6 Million yrs. Ago	?	Archaic Fauna	
MIOCENE		14 Million yrs. Ago	Ramapithecus Kenyapithecus		Modern Flora Evident

FIGURE 2.6

3 Fossilization and Faunal Analysis

FOSSILS AND FOSSIL MATERIALS

In attempting to reconstruct prehistoric ecosystems, paleoanthropologists find that fossilized materials themselves contain considerable information. Once we have the fossils, speculations and hypotheses concerning the existence of certain evolutionary connections cease, since the fossils are the visible evidence of past organisms. As a matter of fact, the fundamental basis of paleontology—the study of ancient life—rests on fossils. By studying the interrelationships between several fossil populations from a single deposit, we often get clues helpful in deciphering the conditions of life that once existed. A fossil, in general, shows the same characteristics as the organism when alive. The location of a fossiliferous bed, plus its associated materials, thus throws light on the paleogeography of the area. And through the use of *radiometry* and *stratigraphy*, a fossil assemblage may be dated (and hence tied into an absolute chronology). Since fossil assemblages and the deposits in which they occur often indicate the nature of the adaptation between the environment and the organisms represented by the fossils, fossils are the best sources of information on evolutionary processes in the past.

When a fossil is unearthed, the question of how to use it in interpreting past events inevitably arises. When such a question arises, we use the present as the key to the past. Where similar and comparable organisms are living today, the ecology and the type of adaptations now seen were probably prevalent during the past times. This is the principle of *uniformitarianism* as applied to paleoecology. But this concept should be applied reasonably; one should take care to see that ecological and chronological factors affecting reconstruction are carefully considered before applying this principle.

What becomes fossilized? Almost any living materials have a chance of becoming fossilized if deposited in the right medium. Some parts of an organism, the hard parts, for instance, will of course have a better chance of getting preserved than others. An organism or its parts must be completely buried or covered by the earth before the onset of organic decay to become fossilized. Dismembered parts, for example, may become buried, or the process of dismemberment and disarticulation of a whole organism may take place after burial. Once the part in question is buried, the organic materials are eventually replaced by inorganic materials (minerals replace tissue, for instance) in the medium in which they are deposited: fossilization thus occurs. Fossiliferous beds are often exposed by erosion.

However, when no erosion takes place, the fossils may never be found!

Certain parts of the skeleton are preserved better than other parts. For example, the teeth, made up of hard enamel, are already in such a condition that preservation takes place immediately. The ends of long bones and small wrist and ankle bones are preserved quite well because of their sturdy nature. Other parts do not preserve so well; for example, the sternum of the skeleton and the softer cartilaginous parts are not readily preserved.

The preservation and fossilization of teeth is of great advantage to the paleoanthropologist. Such preservation led to Davidson Black's discovery of hominid materials in China. In Chinese apothecary shops, fossilized animal teeth have traditionally been ground into powder and used as medicine. Black, then a professor of anatomy at Peking Medical College, systematically looked at these shop wares, found hominid teeth, and established the occurrence of earlier species of *Homo* in China. In addition, teeth are excellent clues for determining the paleoecology of an area, since the teeth of larger mammals provide information on the type of vegetation eaten by these animals.

FAUNAL ANALYSIS

Since fossilization is fragmentary in general, and complete skeletons are rarely preserved in a single deposit, we must use whatever fragmentary materials are available. Three possible approaches exist for determining the number of individuals at a site. One approach involves counting all of the bones of a particular kind of animal in a deposit and then reporting the total number of animals of that type present in the deposit. A second method is to take a diagnostic and highly preserved part (perhaps the limb bones), count only these, left-sided from right-sided bones, sort and then calculate the minimum number of individuals involved. A third approach includes calculating the numerical distribution of materials in a deposit; the emphasis here is to observe the patterns of distribution. Such patterns often indicate the nature of preservation, and, when samples are large enough, give a clue to the approximate numbers of animals involved.

As an example, take the following hypothetical case—the distribution of bone materials for a certain mammal at three sites.

Mammal Type: X

	site A	site B	site C
femur (left and right)	22	14	40
humerus (left and right)	22	9	22
skull	3	8	12
lower jaw bits (left and right)	4	8	13
metapodials	24	16	19

On the basis of distributional patterns, one could argue for the following: at site A around 10 individuals are represented; at site B approximately 8; and at site C some 12 individuals were present in the deposit. In this example the numerical distribution patterns are more significant than the minimum or maximum numbers of bones. This method does have problems; e.g., in nonhuman skeletal material varying butchering techniques used by prehistoric populations may have lead to differential preservation of certain bony parts.

4 Early Hominids and Their Ecology

Our knowledge of early hominid (australopithecine) evolution (figures 4.1 and 4.2) is known best from African finds. From Asia, both *Ramapithecus* (from India) and "Meganthropus palaeojavanicus" (from Java) can be considered as early hominids; however, their ecologies are known only fragmentarily.

This section contains three examples of reconstructed hominid ecologies. Two of them are from Africa (Olduvai Gorge and the Omo Valley region) and one from Asia (the Siwalik Hills of India). Further discoveries in these regions and elsewhere will undoubtedly modify any conclusions. At this time these ideas are meant to "organize our thinking and direct our questions; it remains for archaeologists and others in the behavioral sciences to try to put them in the form of questions that can be answered by explicit research programs."[1]

OLDUVAI GORGE

Africa has played a great role in the study of the early evolution of hominids, particularly through some of the sites in the Olduvai Gorge.[2] A landmark located in the East African Rift Valley in Tanzania, the gorge was discovered by Katt Winklez in the early 1900s while collecting butterflies on an entomological field trip. In 1913 Hans Reck worked there briefly and a period of inactivity followed until 1930. Since the 1930s the indefatigable Leakey family (the late L.S.B. Leakey, his wife Mary, and son Richard) have worked at Olduvai, and have made spectacular discoveries there. The gorge has five stratigraphic beds (I through V), with the lowest (Bed I and the lower part of Bed II) yielding pebble tools and australopithecine fossils along with Villafranchian fauna. Bed I is thick (often more than 120 feet in thickness), and contains tuffs and clays. Animal fossils include those of crocodiles, fish, and mammals. Some of these fossil-bearing beds were overlain by volcanic tuff, sealing over many exposed areas. The Leakeys have reported a living floor (paleosol) horizon of an australopithecine ("Homo habilis"). Bed II is subdivided into a lower and an upper part, the part of bed II contains lacustrine (lake-associated) and fluviatile (river-associated) deposits, and at the top of the lower part of Bed II the Villafranchian fauna clearly ends.

1. G. Hole and R. F. Heizer, *An Introduction to Prehistoric Archaeology* (New York: Holt, Rinehart, and Winston, 1967).

2. L. S. B. Leakey, *Olduvai Gorge I* (Cambridge: Cambridge University Press, 1965); P. Tobias, *Olduvai Gorge II* (Cambridge: Cambridge University Press, 1967).

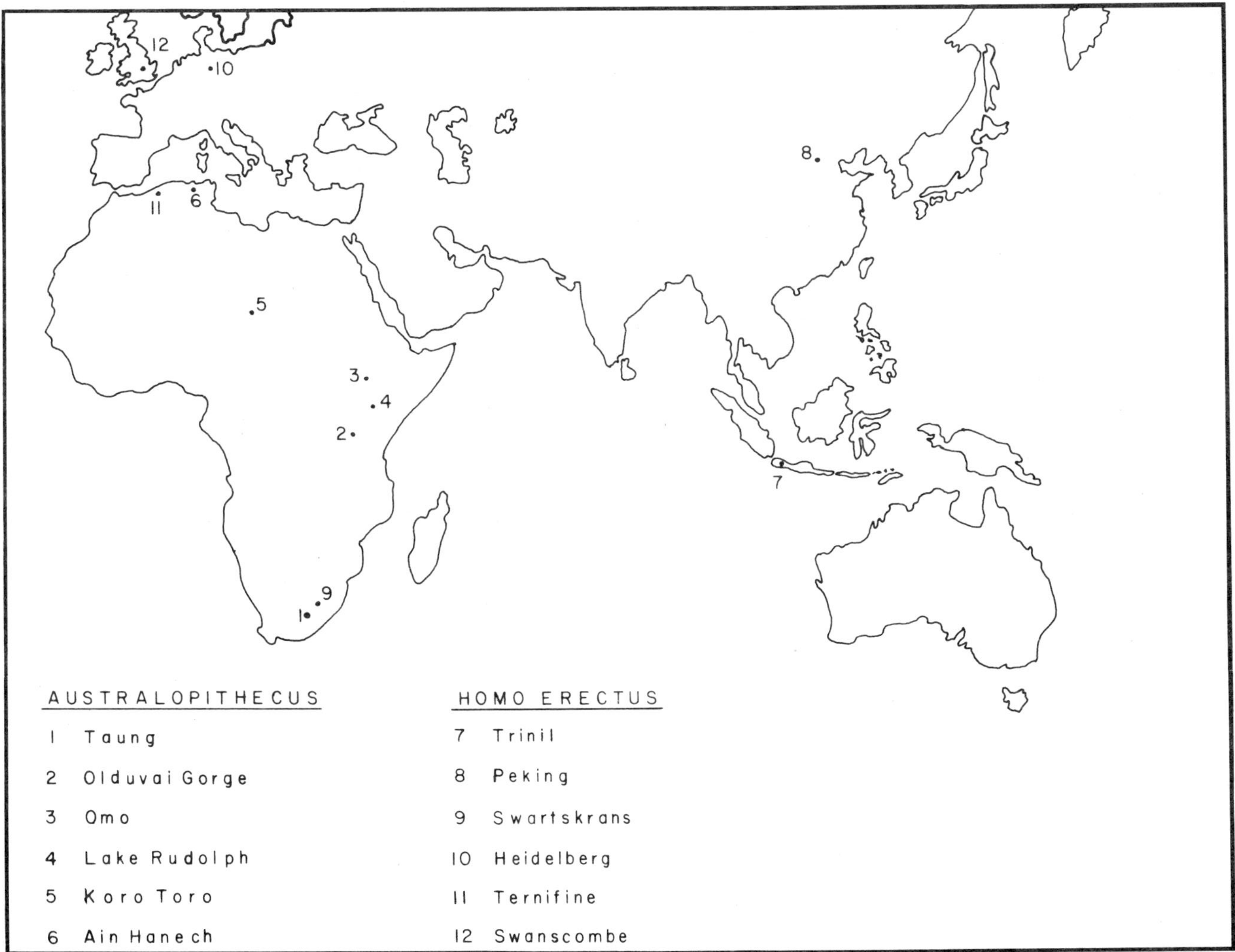

FIGURE 4.1 Select hominid fossiliferous sites.

In Bed I several samples have been analyzed by the K-Ar dating technique, samples chosen to reflect the time period of some of the earliest hominids (australopithecines, Villafranchian fauna, and Oldowan stone tools).

Bottom of Bed I

sample no:	date: (in million years)
KA 412	1.63
KA 437	1.74
KA 846	1.57
KA 847	1.85
KA 849	1.89
KA 850	1.78
KA 851	1.64

Top of Bed II

sample no:	date: (in million years)
KA 664	1.02
KA 664R	1.13
KA 861	1.38

Range of dates for Bed I: 1.6 to 1.9 million years (average = 1.75 million years).

At Olduvai Gorge, occupation by the earliest hominids took place for the most part at the edge of shallow water lakes and swamp land. In Bed I a moderately developed elephant *(Elephas recki),* another proboscidean *(Deinotherium), Rhinoceros,* horses *(Equus* and *Stylohipparion),* a small giraffe, bovids, a carnivore *(Machairodus),* a pig *(Mesochoerus),* and an ancestral type of wart hog *(Phaeocochoerus antiquus)* are all reported. There is a faunal break after the early deposits, such that by upper Bed II a wide variety of pigs and advanced elephants are present; the types of *Rhinoceros* and bovids are different, *Equus* and *Stylohipparion* are still present, but *Deinotherium* is absent. Fossils of elephants and suids (pigs) are extremely useful time markers because they range throughout the main fossil provinces of the continent.

In the lower deposits (Bed I and lower part of Bed II) the fauna indicate a rather moist, tropical woodland and savanna environment. However, this area now receives less than 20 inches annual rainfall, giving it a semidesert character. All in all, the ecological setting of Olduvai's early hominids seems to have been far different from the present-day ecology of the gorge.

OMO RIVER REGION

Located in southwest Ethiopia, the Omo River region, a rich zone, was not extensively explored until the 1960s. Rudimentary study of the region, however, had been done in the past. In 1896, for example, the ill-fated Bottego expedition observed sedimentary rocks and beds of the area without locating the fossiliferous beds. In 1902 E. Brümpt discovered the fossiliferous beds and shipped some fossils to Paris. In 1933 Camille Arambourg did pioneering work sorting out the geology of the area. Since 1967, several international teams (composed of scientists from Kenya, France, and the United States) have been working in the area. Like the Olduvai results the Omo findings have been spectacular. The lower Omo Valley is a tectonic depression, being an extension of the Lake Rudolf trough. Some of the beds have been K-Ar dated; ages range from 1.81 to 4.25 million years.[3]

Today, the plains of the lower Omo Basin have a semiarid, tropical climate, with annual rainfall around the 350 to 600 mm range (approximately 15 to 25 inches). Precipitation is mainly in the form of thundershowers during March through June. Associated with the beds yielding hominid fossils were forest-dwelling animals, such as elephants *(Elephas recki), Deinotherium,* white and

3. K. W. Butzer, "The Lower Omo Basin: Geology Fauna and Hominids of Plio-Pleistocene Formations," *Naturwissen-Schaften* 58(1971):7-16; *Environment and Archaeology* (Chicago: Aldine Publishing Co., 1969).

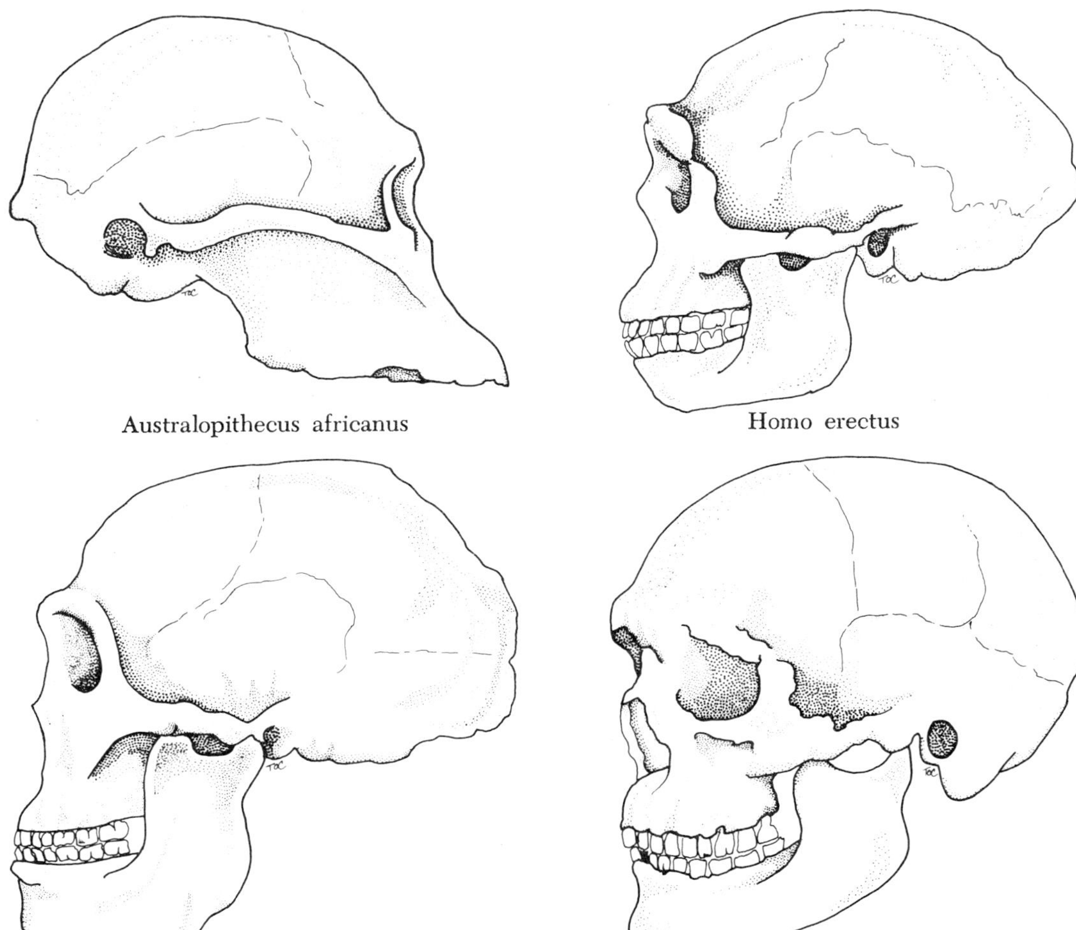

FIGURE 4.2 Early Hominid Types.

black rhinoceroses *(Ceratotherium simum efficase* and *Diceros sicornis),* a camel *(Camelus), Hippopotamus protamphibius,* pigs *(Nyanzochoerus, Pronotochoerus jacksoni),* giraffes *(Libyatherium, Giraffa gracilis),* antelopes, gazelles, impalas, and primates *(Colobus, Cercopithecus, Parapapio, Papio).* The dates for the Omo Valley hominids—australopethecines—extend from 2.5 million years to around 3.5 million years. From fauna and pollen analysis it appears that the Omo region was a "mosaic of grassland, tree savanna, riverine and wetland vegetation."[4]

SIWALIK HILLS REGION

Ramapithecus, from the Siwalik Hills in northern India, is perhaps the earliest hom-

4. Butzer, "The Lower Omo Basin."

inid fossil genus known to us.[5] These fossils have been dated to the Miocene epoch (14 million years ago). The Siwaliks are located in the Outer Himalayas, about five to thirty miles wide and 4,000-9,000 feet high.[6] The geology of the region indicates the following five time periods: Murree, Kamlial, Chinji, Nagri, and Dhok Pathan. The *Ramapithecus* materials come from Chinji and Nagri deposits. The Chinji fauna consists of *Dinotheres,* primitive trilophodonts, forest suids, *Gyraffokeryx* ("water-loving animal"), water deer, crocodiles, pythons, monitor lizards, turtles, and aquatic birds.[7] The hooves of the hoofed-animals are broad, and the dentition is adapted for eating soft, succulent vegetation.[8] The Nagri fauna is similar, and contains *Hipparion,* a marker for the Mio-Pliocene boundary as well as a marsh dweller.[9] Flora remains indicate an abundance of angiosperms, and the evolution of plants of the modern type seems to have been complete by Miocene times.[10]

Some general ideas can be drawn from the African and Asian data presented. These are at best hypotheses to be verified or modified by further work. First, it seems that a mosaic of wet forested and savanna-woodland conditions is associated with the earliest hominids. We may speculate that these conditions probably induced some kind of ecological pressure on hominid populations. In modern contexts, apes in similar ecological conditions are knuckle-walkers; hence, it may be that the earliest hominids did have a knuckle-walking stage before completely becoming erect.[11] However, remember that extrapolating the modern ecological conditions onto prehistoric situations is a doubtful procedure; sometimes the differences between modern and ancient situations are quite striking.

For Further Reading

Bishop, W. W. and G. R. Chapman. "Early Pliocene Sediments and Fossils from the Northern Kenya Rift Valley." *Nature* 26 (1970). Describes Baringo materials within a nine- to twelve-million-year date range.

Casteel, R. W. "Some Archaeological Uses of Fish Remains." *American Antiquity,* 37 no. 3 (1972):404-9. Paper deals with determination of age of a fish at death and the time of year or season it died (and hopefully was deposited at a site) by studying the annual growth marks (annuli).

Chaplin, R. E. *The Study of Animal Bones from Archaeological Sites.* New York: Seminar Press. A brief book that deals with archaeological aspects and faunal analysis.

Clark, J.G.D. *Starr Carr: A Case Study in Bioarchaeology,* Menlo Park, Calif.: Cummings Publishing Co., 1972, 1-42. The physical environment was reconstructed on the basis of pollen, faunal, floral and geological evidences. By use of ethnographic analogy, settlement patterns, locational analysis of artifacts, etc., the materials are interpreted.

Maglio, V. J. "Vertebrate Faunas and Chronology of Hominid-Bearing Sediments East of Lake Rudolph." *Nature* 239 (1972). A com-

5. E. L. Simons, *Primate Evolution* (New York: Macmillan Co., 1972).

6. J. R. Wadia, *Geology of India* (London: Macmillan & Co., 1953).

7. Edwin Colbert, "Siwalik Mammals in the American Museum of Natural History," *Transactions of the American Philosophical Society* 26 (1935):1-401.

8. Ian Tattersal, "Ecology of North Indian *Ramapithecus*." *Nature* 221 (1969).

9. E. Simons; D. Pilbeam; and S. J. Boyer; "Appearance of *Hipparion* in the Tertiary of the Siwalik Hills of North India, Kashmir and West Pakistan," *Nature* 229 (1971).

10. Wadia, *Geology of India;* E. B. Leopold, "Late-Cenozoic Patterns of Plant Extinction," in *Pleistocene Extinctions,* P. S. Martin and H. E. Wright, Jr., eds. (New Haven: Yale University Press, 1967), pp. 203-46.

11. S. L. Washburn and R. Moore, *Ape Into Man* (New York: Little, Brown & Co., 1974).

prehensive description of Lake Rudolph fauna.

Bibliography

Brothwell, D. R. 1965. *Digging up Bones.* London: British Museum.

Schmidt, E. 1973. *Knochenatlas.* New York: John C. Wiley and Sons.

Ziegler, Alan C. 1973. *Inferences from Prehistoric Faunal Remains.* Module in Anthropology, No. 43. Reading, Mass.: Addison-Wesley Publishing Co.

5 Holocene Paleoecology of Ecuador: An Analysis

Recent research on the Santa Elena Peninsula in southwestern Ecuador has brought to light several displacements that have occurred in the Peru Current and the El Niño Current. The Santa Elena Peninsula lies from 2° to 3° south latitude and about 81° west longitude. It is in Guayas Province, and includes part of Santa Elena and Salinas Cantons. The point of Santa Elena (83° 20′ x 2° 11′) forms the westernmost tip of the country. Along the peninsula proper are four marine terraces at about 3 m, 10 m, 20-30 m, and 100 m elevation. The 3 m terrace is of eustatic nature. The 10 m and the 20-30 m terraces are archaeologically significant. The terraces of the peninsula are uniformly overlain by Quaternary deposits in the form of raised beaches, a condition most possibly due to past heavy rainfall.[1] This region is a rising coastline.[2] Very little continental slope exists there.

PACIFIC COASTAL VEGETATION

Throughout the Pacific tropical coasts of South America, the mangrove vegetation is one of the chief characteristics. This vegetation generally extends from the Gulf of Panama to Tumbez in northern Peru, but it forms three major but discontinuous bands. The growth is thick and dense in regions to the north of the equator, particularly from Panama to northern Ecuador. Growth is somewhat less dense in the regions to the south of the equator with the single exception of the Gulf of Guayaquil.[3]

Mangrove occurs in tidal zones on the east and west sides of the South American continent, as well as in the Caribbean area. Mangrove particularly flourishes in the humid tropics. The Colombian sections of Pacific mangrove receive from 200 inches to 400 inches of annual rainfall. There are over 20 species of plants in mangrove, but three of them occur in the three regions previously cited. These are *Rhizophora mangle* (family Rhizophoraceae), *Languncularia racemosa* (family Combretaceae) and *Avicennia nitida* (family Verbenaceae). A fourth form belonging to the genus *Conocorpus* (popularly called buttonwood) sometimes occurs along with the above species. All four share a common trait—a tolerance for salinity. Their climatic and physical requirements are generally uniform. They do not tolerate cold below 20°C nor a gradual deposition of fine alluvial soil.

1. G. W. Sheppard, *The Geology of Southwestern Ecuador* (T. Murby and Co., 1937).

2. V. Oppenheim, "The Structure of Ecuador," *American Journal of Science* 248 (1950):527-39.

3. R. C. West, "Mangrove Swamps of the Pacific Coasts," *Annals of the Association of American Geographers* 46 (1956):98-121.

A typical mangrove forest supports a rich variety of other flora and fauna. Flora are usually composed of algae (collectively called Bostrychidae); fauna consists of snails, bivalves, and snakes. Some bivalves are specific to mangrove vegetation. One such form is *Anadara tuberculosa.* In Ecuador and Peru, these forms are locally called *concha prieta.* In older literature this form has also been described as *Arca tuberculosa.*

R. E. Coker was engaged in making mollusk collections for the Ministerio de Fomento in Peru. He is quoted as follows:

> Concha prieta, Mouth of river Tumbez, and near Capon, from muddy floors of mangrove swamps. Among the first phenomena to catch one's attention on entering the mangrove swamps is a sound, heard repeatedly on every side, as of nuts falling in water or soft mud. Tracing the sound with some care, it is found to come from the watery hollows in the mud occupied by concha prieta, and is presumably made by the sudden closing of the valves under the water by the mollusc (as quoted in Dall, 1910).[4]

In addition to *Anadara tuberculosa,* certain forms belonging to *Cerithidea, Ostreae,*

4. W. H. Dall, "Report on a Collection of Shells from Peru with a Summary of Littoral Marine Molluscs of the Peruvian Zoological Province," *Proceedings of the U.S. National Museum* 1704 (1910): 37.

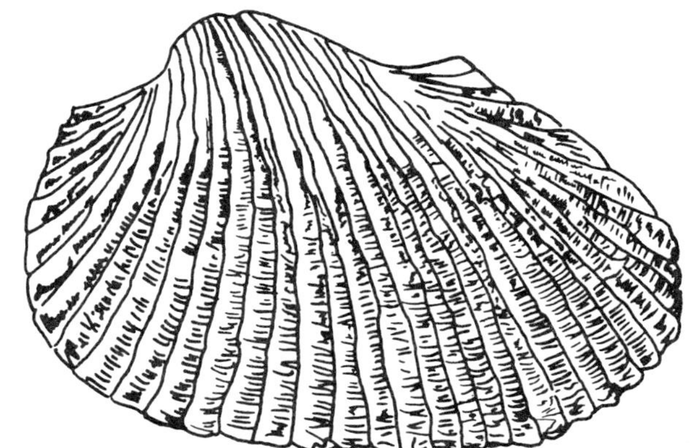

Anadara (Anadara) Tuberculosa (Sowerby 1833) 1:1

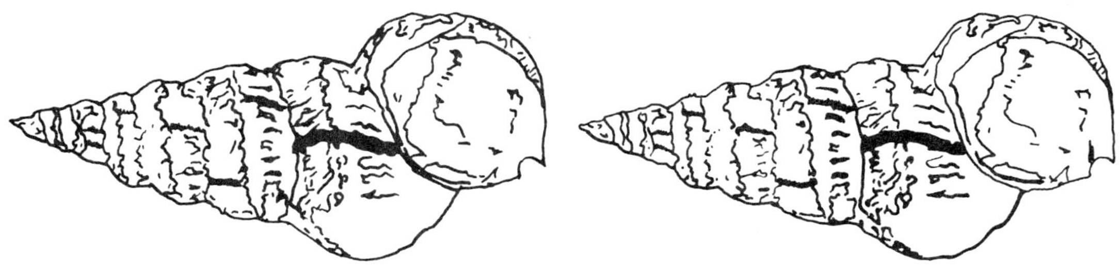

Cerithidea Pulchra (C. B. Adams 1852) × 5

FIGURE 5.1 Mangrove specific habitat dwellers. These animals can give excellent clues to the past ecological conditions.

and *Strombidae* also exist in the mangrove environment, but when such environment is unavailable, they can survive in mudflats. Mangrove swamps have a soil cover of black mud and peat. In the midst of this covering are small isolated sandy patches or spots called *firmes,* rising slightly over the mucky swamp. In Colombia these spots are inhabited by humans. Interestingly, usable fresh water is generally available below the *firmes* at depths of four to five feet, making the growth of coconut palms, maize, and other crops possible in the area. Archaeological objects have been observed by West in the *firmes* near Paita, Tapajo, and Sanguina (all in Colombia).[5]

EFFECTS OF OCEAN CURRENTS

The Pacific coast of South America responds to the effects of both cold and warm ocean currents. In the Santa Elena Peninsula, the effects of the shifts of these currents are fairly pronounced, causing either wet and dry conditions. As the cold Antarctic Current strikes the lower Chilean coast it splits into a northern and southern half. The northern half is driven by southeast winds and runs along the coasts of South America as the Humboldt (or Peru) current. The prevailing southeasterly winds carry the surface water along the current further away from the shore, thus leading to an upwelling of much colder water from below.[6]

The Peru Current generally follows the western coast of the continent in a northerly and north-northeasterly path. However, wind action varies this direction slightly at times. The geographic nature of the coast is such that land juts into the ocean increasingly from south to north. Because of this, the current stays close to the coast until the Santa Elena Peninsula. Thereafter, the land retreats, and the current is diverted away from the land masses. The width of the current is about 120 miles at latitude 33° (near Valparaiso). It has been observed to have the following average temperatures: 39°E at 47° S; 52°F at 33°S; 57°F at 30°S; 64°F at 18° 30′ S (near Arica). The current's average velocity is about fifteen miles per day but falls to about a knot between Arica (18° 30′ S) and Pisco (13° 30′ S), displaying a westerly set at times. The greatest strength is attained between Paita (5° S) and the Galapagos Islands where ships have been known to veer fifty miles west-northwesterly in twenty-four hours. Near Cabo Blanco, the Peru current shifts in a westerly direction towards the Galapagos Islands, where it joins the equatorial current. The surface waters of the Peru current are relatively low in temperature close to shore, with a progressive rise from the shore. Surface temperatures of the water average from 58°F to 64°F. On an average, the temperatures are about 18°F lower than the latitude alone would indicate. Recent studies indicate that the upwelling of the current is limited to a depth of about 100 to 150 fathoms. The Peru Current swings away from land at about 5° S in northern Peru and southern Ecuador. A small branch of the main current, however, continues northward along the coast of Ecuador nearly to the equator, with the result that the coastal regions have a semiarid character.[7]

EL NIÑO: RECENT OBSERVATION ON AN OCEANOGRAPHIC CONDITION

The equatorial countercurrent normally turns counterclockwise along the coasts of Ecuador, but seasonally swings south during the January-March months. Conventionally,

5. West, "Mangrove Swamps of the Pacific Coast." pp. 98-121.

6. W. G. Kendrew, *Climates of the Continents* (Oxford: Oxford University Press, 1963), p. 608.

7. Ibid.

this event coincided with Christmas period, and hence this got named El Niño, or "Baby Christ." During 1971, the seasonal displacement dumped more than 566.5 mm (212 inches) of rainfall along the peninsula. During 1972 a similar phenomena took place, with 315.4 mm (118 inches of rain falling) in an area where the annual average rainfall (over a 40-year period) is less than 100 mm. In some years (1968, for example), no rainfall occurs at all. During 1973, conditions reverted back to the normal rainfall in the peninsula but oceanographically, the warm waters still prevailed along the southwest coasts of Ecuador and northern Peru. The phenomena of 1971, 1972, and 1973 suggested that meteorological changes affecting the region appeared to stretch beyond the limits of an annual period. In other words, those meteorological conditions that were thought to span within a 12-month period, apparently stretched beyond the "annual" time, and continued into the following period. Observations made by Kukla and Kukla for expansion and contraction of pack ice in the northern hemisphere suggests that an extended expansion shifts the oceanic currents southward by several degrees, affecting vegetation, rainfall, and temperature.[8] The rainfall pattern of Southwestern Ecuador supports such a conclusion. During these southward incursions, the warm waters produce high mortality of littoral invertebrates, fishes, and birds. Extensive discoloration of water due to planktonic disturbances (brought or by the warmer temperatures) may also occur. This phenomena is always accompanied by marked changes in sea fauna, including the disappearance of typical cool water birds and the invasion of the region by equatorial waters. The severe Peruvian anchovy losses of 1971 and 1972, as a case in point, triggered a regional food crisis; the losses were caused by this phenomenon.[9]

The southward shift of the equatorial currents, including the El Niño condition, is of importance for the production of wet or dry conditions in southern Ecuador and northern Peru. A more detailed examination of this is given later in this chapter.

ARCHAEOLOGICAL FINDINGS

The peninsula was intensively surveyed during 1964, 1968, and 1971 for archaeological finds. During the survey it was noticed that there were both preceramic and ceramic occupations. A chronological chart is given in figure 5.2. This figure is based on a large number of carbon-14 dates, seriation, and stratigraphy.[10] The earliest occupations are the preceramic Vegas industry; the ceramic industries then follow. The ceramic occupations of the Santa Elena Peninsula are well-known, having been under study for a long time. These are (from earliest to latest) Valdivia, Machalilla, Engoroy, Guangala, and Libertad. Valdivia pottery has been claimed to be the earliest in the

8. G. Kukla and H. Kukla, "Increased Surface Albedo in the Northern Hemisphere," *Science* 4126 (1974):709-14.

9. Each time the southward invasion of warm water takes place near the Ecuadoran and Peruvian coasts, a biological catastrophe occurs; all the anchovy, a local food staple, disappear. Due to the rapid anchovy kill, hydrogen sulfide and its accompanying putrid smell are evident in the locale.

10. Edward P. Lanning, *Peru Before the Incas* (Englewood Cliffs, N.J.: Prentice-Hall, 1967); Akkaraju Sarma, "The Cultural Implications of Upper Pleistocene and Holocene Ecology of Santa Elena Peninsula," (Ph.D. diss., Columbia University, 1969); E. J. McDougle, "Water Use and Settlements in the Changing Environments of the Southern Ecuadoran Coast," (Master's thesis, Columbia University, 1967): A. C. Paulsen, "Chronology of Guangala and Libertad Ceramics of the Santa Elena Peninsula in South Coastal Ecuador," (Ph.D. diss., Columbia University, 1970).

Ceramic Period	Limit Dates Established on Stylistic Basis, C-14 Dates and Stratigraphic Grounds.
Libertad 6	1475 - 1532
Libertad 5	1400 - 1475
Libertad 4	1324 - 1400
Libertad 3	1250 - 1324
Libertad 2	1175 - 1250
Libertad 1	1100 - 1175
No Evidences of Occupation for 800-1100 A.D.	
Guangala 8	750 - 800
Guangala 7	650 - 750
Guangala 6	600 - 650
Guangala 5	500 - 600
Guangala 4	450 - 500
Guangala 3	200 - 450
Guangala 2	100 - 200
Guangala 1	BC 100 - 100 AD
Engoroy	850 - 200
Machalilla	1000 - 850
No Evidences of Occupation for 1600-1000 B.C.	
Valdivia 8	1700 - 1600
Valdivia 7	1800 - 1700
Valdivia 6	1900 - 1800
Valdivia 5	2000 - 1900
Valdivia 4	2100 - 2000
Valdivia 3	2300 - 2100
Valdivia 2	2400 - 2300
Valdivia 1	2400 - 2650
No Evidences of Occupation for 5000-2650 B.C.	
Preceramic Period	
Vegas	6800 - 5000

FIGURE 5.2

Americas,[11] but this has been questioned on stylistic grounds.[12]

The peninsula was unoccupied at certain times; these periods of abandonment reflect arid times. Since water is a critical variable for human occupation, it has been postulated that during arid times the peninsula was either abandoned or very sparsely occupied.[13]

FINDINGS FROM THE ANALYSIS OF MOLLUSKS

The shell middens of the peninsula contained over 40 different species, which were analyzed in terms of contrasting habitats—mangrove dwellers versus intertidal dwellers, for example. As pointed out earlier, mangrove vegetation needs a large amount of rainfall and silt. And the presence of mangrove forms indicated the availability of water in the form of rainfall. Therefore the relative presence or absence of mangrove-dwelling mollusks gives the pluvial patterns of the region. The data obtained from shell midden analysis is presented in graph form (see Figure 5.3), with time on one axis and the percentages of *Anadara tuberculosa* specimens and all similar habitat dwellers on the other axis. The two lines denoting the percentages on the graph represent the figures for all wet markers.

From the graph, it appears that certain periods were suitable for mangrove vegetation, as evidenced from increases in the numbers of mangrove-dwelling mollusks. These increases in number must also have been during periods of increased pluviality.

What caused pluvial conditions in the peninsular region? It has been suggested that rainfall is caused in the peninsula by a

11. B. J. Meggers; C. Evans; and E. Estrada, *Early Formative Period of Coastal Ecuador: the Valdivia and Machalilla Phases*. Smithsonian Contributions to Anthropology, no. 1 (Washington, D.C.: Smithsonian Institution, 1965), pp. 1-234.

12. D. Collier, review of B. J. Megger's "Ecuador," in *American Antiquities* 33 (1968):267-71; Edward P. Lanning, "Archaeological Investigations on the Santa Elena Peninsula, Ecuador," Report to the National Science Foundation on Research Carried Out under Grant GS-402, 1964-1965 (New York: Columbia University, 1967, mimeographed); I. Shimada, "An Evaluation of the Validity of Meggers-Evans's Trans-Pacific Cultural Contact Hypothesis and Possible Alternative Hypothesis," (manuscript in the Department of Anthropology, Cornell University, Ithaca, New York, 1970).

13. A. Sarma, "Holocene Paleoecology of South Coastal Ecuador," *Proceedings of the American Philosophical Society* 118, no. 1 (1974):93-149; McDougle, "Water Use and Settlements in Changing Environments of Southern Ecuadorean Coast" (Master's Thesis, Columbia University, 1967).

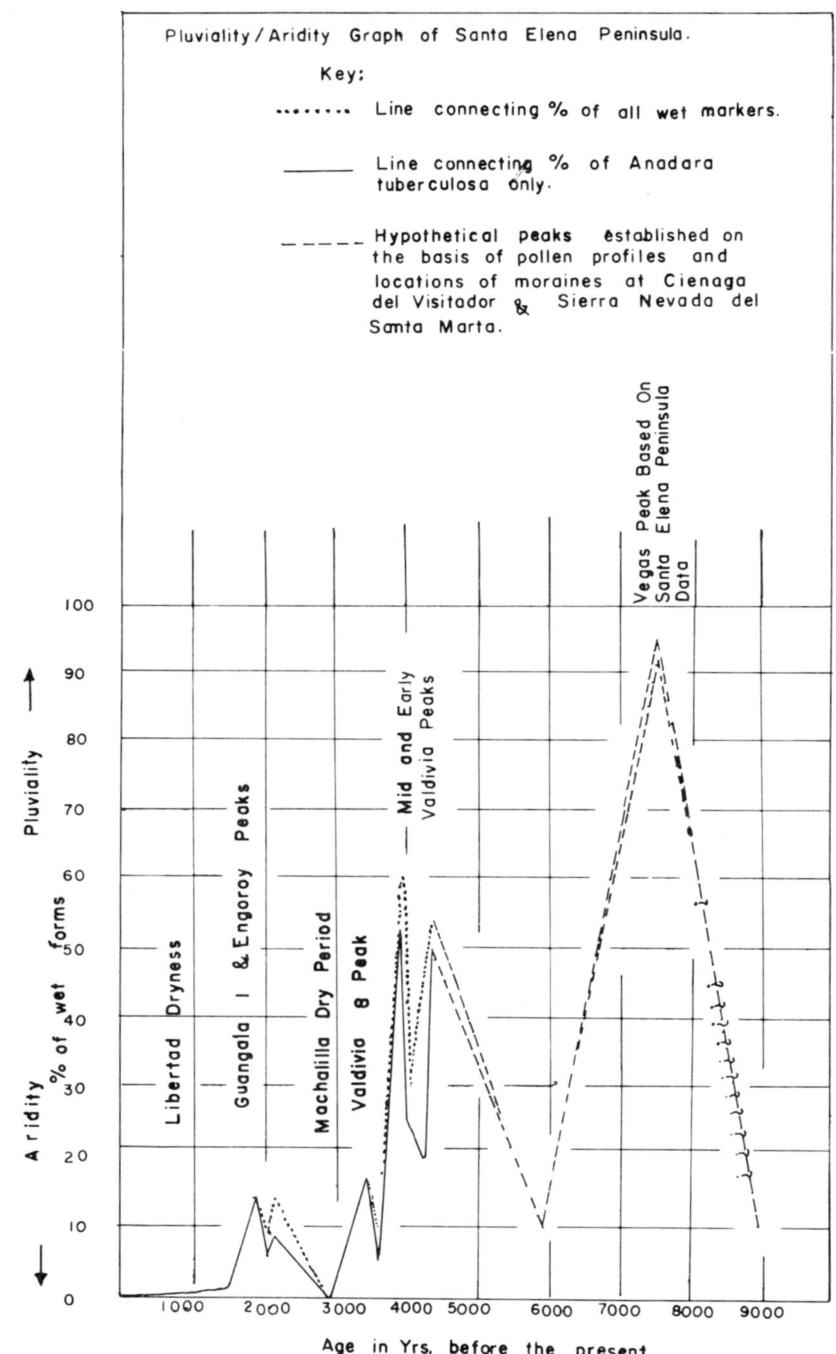
Pluviality / Aridity Graph of Santa Elena Peninsula.
Key:
Line connecting % of all wet markers.
Line connecting % of Anadara tuberculosa only.
Hypothetical peaks established on the basis of pollen profiles and locations of moraines at Cienaga del Visitador & Sierra Nevada del Santa Marta.
Vegas Peak Based On Santa Elena Peninsula Data
Mid and Early Valdivia Peaks
Libertad Dryness
Guangala I & Engoroy Peaks
Machalilla Dry Period
Valdivia 8 Peak
Aridity % of wet forms Pluviality
100
90
80
70
60
50
40
30
20
10
0
1000
2000
3000
4000
5000
6000
7000
8000
9000
Age in Yrs. before the present

FIGURE 5.3

seven-year cycle (based on historical records), although evidence demonstrates that the cycle skips occasionally.[14] Gerhard Schott suggested a more appropriate explanation—a southward shift of the rain-causing front called by him the "meteorological equator," also known as the *intertropical front* (ITF).[15] This front may once have been much further south than its present summer and winter locations. Archaeological evidences from Santa Elena Peninsula and the Talara region of Peru (the Siches complex) show that the mangrove was dense at times coinciding with pluviality on the graph.[16] Therefore it is likely that a shift of the front extended as far south as the Talara region (5° S). This shift of the front is being part of larger dynamic changes: the shift of the front also moved the Peru Current farther south than the present location; and the El Niño condition and equatorial currents shifted southward too. This shift also produced shifts in the climatic bands and vegetational zones. Interestingly, a similar degree of latitudinal changes in climate-vegetation zones occurred in Africa.[17]

Speculating on the 1971-72 rainfall patterns and pack-ice explanations, it has already been pointed out that ice pack expansions shift all dynamic meteorological factors southward. Therefore in the extended stadial times, these shifts were more prolonged, a concomitant cold pluvial condition could support extensive mangrove at the peninsular region.

From figure 2.1, observe that the southward displacement was extended in time. The archaeological evidence (at least on the basis of our present knowledge of the peninsula) shows that during Vegas 6800-5000 B.C.), Valdivia I (2650-2450 B.C.), Valdivia 6 through 3 (2100-1800 B.C.) and Guangala (100 B.C.-100 A.D.), there is an increase in the percentages of wet markers, indicating pluviality. These findings correlate well with the known history of pollen in Colombia, British Guiana, and seem to reflect a widespread Andean *cold pluvial* pattern.[18]

Thus it seems that in the past 9,000 years the western coast of South America underwent climatic changes. In such regions as the Santa Elena Peninsula (whose marginal characteristics are ideally suited for paleoecological observations in archaeological contexts), there was a periodic abandonment of the peninsula during arid times, and occupation was reestablished during pluvial times. This periodic abandonment of the peninsula seems to have continued until late Engoroy times. Thereafter the inhabitants attempted to overcome aridity by digging wells, many of which have been dated to Guangala and Libertad times.[19]

The archaeological evidences from south coastal Ecuador support the hypotheses of shifts of the intertropical front southward

14. R. C. Murphy, "Oceanic and Climatic Phenomena Along the West Coast of South America during 1925," *Geographical Review* 43 (1953): 565-6.

15. G. Schott, "The Humboldt Current in Relation to Land and Sea Conditions of the Peruvian Coast," *Geography* 7 (1932):87-9.

16. Sarma, "Cultural Implications;" J. B. Richardson III, "Archaeological Evidence for Post-Pleistocene Climatic Change in Northwestern Peru and Southern Ecuador," (manuscript in the Department of Anthropology, University of Illinois, Urbana, 1966).

17. E. M. Van Zinderen Bakker, "Upper Pleistocene and Holocene Stratigraphy and Ecology on the Basis of Vegetation Changes in Sub-Saharan Africa," in *Background to Evolution in Africa,* W. W. Bishop and J. D. Clark, eds. (Chicago: University of Chicago Press, 1968), pp. 125-147.

18. T. Van der Hammen, "A Palynological Study on the Quaternary of British Guiana," *Leidse Geologische Mededelingen* 29 (1963):125-179; ——— and E. Gonzalez, "Upper Pleistocene and Holocene Climate and Vegetation of the 'Sabana de Bogota,' Colombia, South America," *Leidse Geologische Mededelingen* 25 (1960):261-315.

19. McDougle, "Water Use"; Paulsen, "Chronology."

(along with all the related phenomena) from its present location. Since the archaeological evidences show fairly extended periods of pluviality, it would appear that a seven-year cycle would be inadequate to support mangrove vegetation in any dense matter. These evidences all go to point out extended fluctuations in south coastal Ecuadorian regions of pluviality. The exact relationship of the Peru-Chile undercurrent to the findings at south coastal Ecuador during prehistoric times are unclear still.[20] Perhaps these undercurrents shifted southward at the time of pluvality in south coastal Ecuador.

For Further Reading

U.S. Navy Hydrography Office. *Peru Current: Pilot Chart.* Washington, D.C.: U.S. Government Printing Office, 1963. A descriptive chart.

20. W. S. Wooster and M. Gilmartin, "The Peru-Chile Undercurrent," *Journal of Marine Resources* 19 (1961):97-122.

6 Theoretical Ecology; Paleoecology and Anthropology

THEORETICAL ECOLOGY

Most of the paleoecological observations we have made are based on information from such associated finds as faunal and floral remains, geological data, and other pertinent information. Using data obtained from field studies (from which our conclusions generally are derived), ecologists are now attempting to predict changes as well as generalize laws.[1] In these attempts, computers and mathematical models have served as key aids, with the result that mathematical models underlie much of the basis of theoretical ecology. Here the emphasis has been on the delineation of key components of a system and the isolated treatment of each component as much as possible. One resulting concept is that of *species equilibrium*—a concept applied to conservation problems in national parks, for example. Using an "island" model, for instance, research suggests that the number of species on recently depopulated islands will increase exponentially as species return to "islands," then at equilibrium the number of species that will become extinct will equal those that arrive on the island. Predictions are that the highest rates of extinction will occur among rare species and on small "islands."

The above is one of the several generalizations resulting from theoretical ecology. (This particular model is the MacArthur-Wilson model.) In anthropology, and particularly in paleoanthropology and paleoecology, researchers have not been sensitive to the issues of extinction rates. But in the above light some interesting problems can be raised. For example, when the fauna of some archaeological sites are compared, the animals from those sites that existed later in time are often different from the species represented at earlier sites. It has generally been assumed that the groups of animals at the later sites were hunted by either a new group of people or that cultural changes occurred within the old group that made them concentrate on new prey. We have paid but scant attention to the processes of extinction as "ecological processes." Instead they have hitherto been considered solely as "cultural processes." In paleoecological studies, although the data are limited, the implications of species equilibrium and extinction rates are seminal to our understanding of such issues as megafaunal extinctions, pleistocene overkill, hunting patterns and faunal changes in prehistoric contexts, and the like rather than the simplistic assump-

1. G. B. Kolata, "Theoretical Ecology—Beginnings of a Predictive Science," *Science* 183 (1974): 402-50.

tions we have had thus far. In this discussion theoretical ecological considerations are presented only in passing to show that ecology is not just fieldwork *per se,* but involves generalization, model building, and the development of theory.

Recently, the effects of ecological change have been applied to modern populations as predicted by theoretical models, which in general have fit the observations rather well.[2] In addition, mathematical indices have been devised by biogeographers for comparing populations as to their similarities and differences. They are extremely useful at different levels of archaeological inferences.

PALEOECOLOGY AND ANTHROPOLOGY

The significance of paleoecology for anthropology has only recently been recognized. In recognizing the relationship of paleoecology to anthropology, four major points have been made at one time or another. In brief, these are:

(1) All cultures, prehistoric and recent, exist in an ecological setting.
(2) Cultures must have a biocultural adaptation to survive in such a setting.
(3) A culture is *molded* or *determined* by the characteristics of the environment (i.e., by ecology).
(4) In describing the ecology or environment, modern conditions are taken to reflect the conditions of prehistoric times.

The above points have been stated either collectively or as independent points of view. In evaluating these points, the first and the second ones are generally recognized as valid. The third and fourth points, however, are open to question. As a general rule, in extremely arid areas, like the Australian desert, this environmental determinism may have a general validity; but in other regions the role of the environment as a determinant of culture must be carefully examined.

With respect to the fourth point above, without direct access to prehistoric environments, we may extrapolate the present to the past conditions and develop explicit models.[3] In the development of environmental models one typology suggested in recent years is a four-type system proposed by Meggers in 1954. While Meggers's model is a good one to begin this discussion, two examples from tropical South America will be discussed briefly in the latter part of this section to point out the shortcomings of the model. The various environmental types in the model may be outlined as follows:

Type 1: Areas of No Agricultural Potential. This type includes the greatest variety of natural landscapes. The defective element may be soil composition, temperature, rainfall, short growing season, elevation of terrain, etc. Typical examples: tundras, some deserts, tropical savannas, swamps, some mountain ranges, and similar uncultivable types of lands. Some of the type 1 areas could be pastoral.

Type 2: Areas of Limited Agricultural Potential. Though agriculture can be undertaken here, its productivity is minimized due "to limited soil fertility which cannot be economically improved or conserved." Even when the land is cleared and crops are planted, the delicate balance between the soil in its normal input-output relationships are disturbed. For instance, nutrients are quickly lost by planting. An example of a type 2 area is the South American tropics.

Type 3: Areas of Increasable (Improvable) Agricultural Potential. These areas contain all the essentials of agricultural pro-

2. G. B. Kolata, "Human Biogeography: Similarities Between Man and Beast," *Science* 185 (1974):134-35.

3. B. J. Meggers, "Environmental Limitation on the Development of Culture," *American Anthropologist* 56 (1954):801-23.

duction. Since such areas usually occur in more temperate climates, rainfall and humidity are less detrimental; soil exhaustion is mainly caused by raising food crops. But when slash and burn agriculture is employed, this region is no better than type 2 areas. However, crop yield can increase when farmers rotate their crops, let fields lie fallow, and the like. Examples of type 3 areas are: the temperate forest zones of Europe and of the eastern United States.

Type 4: Areas of Unlimited Agricultural Potential. In these areas, such factors as climate, water, terrain, and soil fertility are ideal for agriculture. Examples are: the Near Eastern "cradles of civilization"; the midwestern United States.

One reason for this classification is that it is "theoretically independent of time factor"; according to Meggers.[4] Since the introduction of agriculture in most of the world, little alteration in either the climate or topography has affected the agricultural potential of the environment. Recently, however, Bryson has shown that climatic changes can occur in a short time and has attempted to explain the reasons for these changes.[5] The literature on climatic changes is quite extensive. Ferdon reevaluated the criteria used by Meggers, by considering the range of variation of precipitation, temperature, soils, and landforms and assigned an alphabetical value to each of these factors.[6] For example, in the case of precipitation, the value A represents rainfall throughout the year; B equals adequate rainfall for six months; C denotes adequate precipitation for highly limited production, and D equals inadequate rainfall throughout the year. In the case of soils, A meant good, B was fair, C represented poor, and D meant no soil. In areas with A type of precipitation and C type of soil, agriculture would be disappointing, no matter the effort nor the amount of rainfall. While Ferdon's refinement is important, he failed to take the time element into consideration. Any attempt to reconstruct an area's ecology without considering time cannot be considered as definitive. As has been pointed out already in the case of early hominids, the paleoecology in areas where the fossils occurred was far different from the present-day conditions. Also, different plant species themselves have variable optimal factors; thus any overgeneralization for all plants can be erroneous.

Some of the recent studies in South America point out the shortcomings of these deterministic approaches. In coastal Brazil, Bigarella and Andrade have studied some recent (Quaternary) sites.[7] According to Meggers's typology this area would be in the type 1 category, as the vegetation cover in coastal Brazil is tropical forest and savanna. In these regions, Bigarella and Andrade observed evidence of climatic changes that alternated between humid and arid conditions. The last climatic fluctuations along the Brazilian coast seem to have ended about 450 B.C., from which time an arid condition prevailed. In the humid epoch, the forest cover reached its maximum expansion, and during the periods of drought the forest retreated and the grasslands spread. In addition, contrary to the usual ideas, these humid epochs in Brazil coincided with interglacial periods (warm times) rather than with glacial (cold) periods. Therefore in certain world regions at least, the oscillation between glacial and interglacial conditions coincided with the decrease and increase in rainfall (or humid-

4. Ibid.

5. Reid Bryson, "A Perspective on Climatic Change," *Science* 4138 (1974):753-60.

6. Edwin Ferdon, Jr., "Agricultural Potential and the Development of Culture," *Southwestern Journal of Anthropology* 15 (1959):1-19.

7. J. J. Bigarella and G. O. Andrade, "Contributions to the Study of the Brazilian Quaternary," *The Geological Society of America,* Special Paper 84 (1965):433-50.

ity), respectively. Whether one is concerned with the rise of civilizations, of agriculture, of states, or whatever, the extrapolation of present-day ecological conditions to prehistoric situations—without verification is invalid.

In yet another study conducted in tropical South America (in tropical savanna of Colombia and Rupununni savannas of Guyana), Wijmstra and Van der Hammen reported in 1966 the alternation between open and closed savannas.[8] These two areas fall under type 1 of Meggers's classification. The open savannas reflect wet conditions with a large predominance of herbaceous plants. The closed savannas have a preponderance of Brysonima plants (essentially woody vegetation), reflecting dry conditions. Several available carbon-14 dates indicate that such alternation of vegetation took place within the last three thousand years or so.

A more recent expansion of deterministic arguments for the Amazon region are expressed by Meggers, who points out that no urban society has evolved there because the environment only produced local variations of a generalized pattern of tropical forest culture.[9] Again, the basic weakness of this and other deterministic arguments is in the extrapolation of the present-day environment to prehistoric conditions. Every effort must be made to collect data on paleoecological conditions and then use the subsequent reconstruction for anthropological inferences.

As a point of contrast, the southeast coastal Ecuadorian and northern Peruvian evidences (see chapter 5) show the extreme importance of the time element. It shows quite clearly that the ecology of the region has fluctuated from one of wetness to one of dryness, leading to the periodic abandonment of region. Clearly, an extrapolation of modern environmental data to the preceramic times (6500-5000 B.C.) in southwest Ecuador is not only meaningless, but incorrect. It is therefore necessary to determine the ecology of the "time" period from which the materials come, rather than substitute a modern, perhaps different, one.

For Further Reading

Hutchinson, G. E. and R. H. MacArthur. "A Theoretical Ecological Model of Size Distribution Among Species of Mammals." *American Naturalist* 93 (1959):117-25. A paper dealing with sizes of animals and their space usage. Large animals use disproportionately more space than small ones; their ranges could be logarithmic in relationship.

Krantz, Grover S. "Human Activities and Megafaunal Extinction." *American Scientist* 58, no. 2 (1970):164-70. The paper discusses changing ecological and animal relationships, as well as extinction factors.

Martin, P. S. and H. Wright. *Pleistocene Extinctions*. New Haven: Yale University Press, 1968. Proceedings of a symposium in which the author argued for the human causes of megafaunal extinctions.

Vayda, A. P. and R. Rappaport. "Ecology, Cultural and Non-Cultural." In *Introduction to Cultural Anthropology*, ed. J. Clifton. Boston: Houghton Mifflin, 1966. An excellent summary of ecological concepts and their applications in anthropology.

Bibliography

MacArthur, R. H. and Connell, J. H. 1964. *The Biology of Populations*. New York: John C. Wiley and Sons.

Netting, Robert McC. 1971. *The Ecological Approach in Cultural Study*. Module in Anthropology, no. 6. Reading, Mass.: Addison-Wesley Publishing Co., pp. 1-30.

8. T. A. Wijmstra and T. Van der Hammen, "Palynological Data on the History of Tropical Savannas in Northern South America," *Leidse Geologische Mededelingen* 38 (1966):71-90.

9. B. J. Meggers, *Amazonia* (Chicago: Aldine Publishing Co., 1971).

7 Use of Modern Climatic Data for Ecological Reconstructions: An Illustrative Case

Much of the details in this section are based on research work done in late 1973, and are being published for the first time.

In the preceding case of Ecuador, it was pointed out that in southwestern Ecuador in 1971-72 rainfall showed a several-fold increase. In the summer of 1973, the archives of the departments of meteorology of Peru, Ecuador, and Colombia were searched, and by correspondence with scholars who had recently done fieldwork in Panama and Guatemala, the annual rainfall variations were studied. The essential purpose of collecting data was to gather the rainfall figures for the last two or three decades from Pacific coastal meteorological stations of all these countries. The primary question was: How far south did the 1971 shift of equatorial currents extend? The 1971 phenomenon characteristically produced increased rainfall along the Pacific coastal stations. The secondary question was: What was the pattern of rainfall for the year 1972 when coastal Ecuador was inundated with rain? Also, data were sought and obtained from a meteorological station (Ancón in Santa Elena Peninsula), whose marginal location would offer clues to the understanding of the seven-year rainfall cycle. The data from other coastal meteorological stations are furnished in Figure 7.1.

The stations in the above table (from top to bottom), are from Meso-America (Guatemala and Panama), and then South America (Colombia, Ecuador, and Peru). An examination of the amount of rainfall for the year 1971 indicates the following. The Meso-American stations show significant drops in the rainfall. This pattern continues on into 1972, when significant 1971-like southward incursions of the equatorial warm currents took place. In the archaeological record, from southwestern Ecuador, such southwardly extended incursions of tropical currents occurred. Also the expansion and contractions of pack ice correlates well with the environmental phenomena of 1971 and 1972.[1] For the Andean regions, evidences clearly point out that a cold pluvial pattern existed.[2] Cold pluvials simply mean that it rained often during a cold phase (during glacial periods, and later stadials, which are extended cold periods within the last

1. G. Kukla and H. Kukla, "Increased Surface Albedo in the Northern Hemisphere," *Science* 4126 (1974):709-14.

2. T. Van der Hammen, "A Palynological Study on the Quaternary of British Guiana," *Leidse Geologische Mededelingen* 29 (1963):125-79; ——— and E. Gonzalez, "Upper Pleistocene and Holocene Climate and Vegetation of the 'Sabana de Bogotá,' Colombia, South America," *Leidse Geologische Mededelingen* 25 (1960):261-315.

FIGURE 7.1 Comparison of recent rainfall from select stations of Western South America (in mm)

Statistical Calculation on the Rainfall Data

	Years	1960	1961	1962	1963	1964	1965	1966	1967	1968
GUATEMALA	Guatemala City	1477.2	1012.1	1302.9	947.5	1238.0	1047.4	1246.9	885.0	1054.7
PANAMA	Cristóbal Colón			3525.5	2777.5	2832.3	3330.6	3580.3	3164.0	2620.0
	Portobelo			4232.7	4416.5	3659.1	4765.2	4922.7	4302.9	3519.9
	Angosturade Cochea						2988.7	4356.4	5273.3	5658.0
	Caldera				3537.1	3476.4	—	—	—	4657.1
	Cerro Punta				1839.3	1767.6	1835.0	—	2066.0	2337.5
	Finca Lerida						1903.7	2634.7	2658.0	3029.5
	Hornito			4023.7	4389.6	—	4412.6	—	4813.8	5032.0
	La Cordillera			—	—	3392.5	2437.1	3854.3	3294.5	4067.5
	Planta Caldera			4345.6	4669.8	4901.8	3022.8	4907.0	4352.3	5065.9
	Potrerillos Arriba			—		4847.4	—	—	4864.0	4855.0
	Darien Boca de Cupe			—	1854.5	—	1746.0	2565.5	2357.5	2349.5
	Los Angles								—	—
	Los Santos						840.5	1025.5	989.0	1364.5
	Tonosi			2270.0	1984.0	—	—	—	—	—
	Chepo			1866.5	2026.5	2678.0	2083.5	2483.5	2299.0	1991.5
	Majé			1809.1	1904.9	—	1935.7	2114.5	—	2077.0
	Tocumen			1479.7	—	1965.7	1813.3	1921.4	1896.4	1799.7
	El Corbrizo			2012.8	2900.1	—	1941.7	3748.1	—	2888.5
	El Palmar			2251.0	2443.5	3477.5	1783.0	2756.0	2199.0	2609.5
	Laguna San Juan			3537.0	3707.3	4460.4	1535.2	3080.2	3288.4	3471.7
	Santa Fe			1818.5	2087.5	—	1495.5	—	2094.0	2355.0
	Sitio del Desvio			—	3407.3	—	2331.4	4175.3	—	4092.8
COLOMBIA	Bahia Salano	3921.0	8320.0	7747.9	3925.1	4626.9	2523.2	2910.6	2354.0	1602.6
	Quibdó	8589.2	8820.0	4603.6	3625.3	7649.0	7278.1	7411.2	7902.1	7399.1
	Buenoventura			4948.3	5424.4	7118.0	5816.6	2538.7	5013.1	6454.1
	Tumaco	2164.0	521.4	676.9	485.3	1895.1	—	1120.7	—	1776.0
ECUADOR	Ancón	47.7	171.6	15.7	13.4	154.6	87.7	25.5	95.5	—
	Salinas	44.5	155.6	8.4	39.2	127.2	116.6	30.2	41.6	8.9
	Manglar Alto			21.2	147.5	351.5	378.1	327.5	847.6	169.3
	Febres Cordero						232.2	331.8	104.0	96.8
	Guayaquil-Salinas Road (Kilometer 28)				120.2	295.7	644.2	190.9	543.3	43.5
	Playas									
	Guayaquil	—	706.6	617.9	597.9	855.4	1322.2	960.9	1014.4	396.6
	Pastaza									
	Cañar									
	Cuenca									
PERU	Talara	0	0	0.2	0	0	0	0	0	0
	Lambayeque	10.0	11.0	—	18.4	—	40.6	14.6	22.9	1.1
	Casa Grande		—	2.6	15.1	21.3	22.1	—	28.4	5.5
	Campo del Marte	18.2	22.2	19.1	35.9	6.2	15.0	7.1	9.8	3.8
	Pisco	3.2	2.4	7.0	12.0	0.5				

1969	1970	1971	1972	Mean(X)	S.D.	S.E.
1671.2	1473.8	1210.1	848.2	1185.769	250.330	69.429
3238.4	4223.9	2881.6		3217.410	479.262	151.556
3754.1	4886.2	3652.0		4211.113	539.974	170.754
5683.3	4520.8	4199.7		4668.600	960.506	363.037
4981.4	4862.3	4283.0		4299.550	658.578	269.863
—	3464.0	2011.5		2840.055	598.836	226.338
3213.0	3827.5	2617.5		2840.557	598.836	226.338
—	6609.1	4421.2		3522.265	526.363	954.875
4568.3	5240.0	4585.0		3929.962	886.406	313.392
6064.9	5273.0	3340.1		4594.400	894.879	282.985
5736.0	5975.0	4964.5		5206.983	509.771	208.113
—	4030.5	3586.0		2064.900	1089.865	411.930
—	1442.5	1332.5		Data inadequate		
1045.5	1601.0	1239.0		1157.857	259.817	98.201
—	2609.3	2159.0		Data too small		
2203.6	3574.5	2228.0		2343.460	494.555	156.392
2096.0	3096.8	1815.0		2106.125	418.325	147.900
1931.3	2153.6	1972.6		1881.522	182.519	60.839
3457.2	4005.9	3221.3		3021.950	750.486	265.337
2737.0	3680.7	2454.8		2639.200	573.625	181.392
3472.4	3868.4	2001.4		3242.240	867.104	274.202
—	3035.5	1828.0		2102.087	493.063	186.360
4311.3	4218.5	3814.6		3764.428	703.010	265.712
1508.0	2987.7	5184.2		3967.600	2206.666	637.009
6647.3	8947.5	3078.7		6829.258	1989.835	574.416
6375.3	6838.0	5738.2		5626.470	1308.014	413.630
2052.2	2404.2	6131.1		1922.690	1641.974	514.238
133.5	117.8	566.7	315.4	145.400	157.235	45.389
—	18.8	299.0	277.0	97.250	101.480	29.295
313.7	359.4	NA	NA	834.791	297.179	85.788
151.1	85.6	356.3	600.4	244.775	178.271	63.028
127.5	92.1	399.9	552.5	300.980	219.700	69.475
339.9				Data inadequate		
8947.0	616.2	659.8	1374.9	834.791	297.179	85.788
		4499.9	4250.7	Data inadequate		
	582.6	532.7		Data inadequate		
	902.6			Data inadequate		
0	1.0	3.7	133.6	9.896	35.619	9.519
12.1	11.2	73.0	81.0	26.900	26.737	8.061
13.8	14.9	120.5	47.6	29.180	34.476	10.902
13.0	30.0	15.2	10.7	15.861	9.335	2.589
				5.020	4.562	2.040

10,000 years), and it was dry during the other times (i.e., interglacials and interstadials). The Ecuadorian prehistoric evidences point out a very good correlation for the Andean cold pluvial pattern.

PALEOECOLOGICAL SIGNIFICANCE

On the basis of 1971 and 1972 evidences, we can speculate about some of the paleoecological implications. The location of tropical high-pressure cells along the southwestern region of the United States produced (along with other environmental phenomena) extreme aridity. Extensive evidence indicates past latitudinal shifts in the vegetational bands due to environmental phenomena. Also, from Figure 7.1, we see an extended southward shift of warm water that seems to have increased rainfall till Campo del Marte (Lima, Peru). Such past prehistoric shifts are on record. Such shifts, on the basis of 1971 and 1972 phenomena, indicate that today's extensive tropical forests of the Gulf of Panama region would have been much less dense, making human crossing of the isthmus region less an ordeal. Anthropologists have often wondered how the first travelers of the isthmus region hacked away the thick jungle with so inadequate tools. When we reconstruct the paleoecology on the basis of a meteorological model, the problem dwindles. The jungle was not very thick.

What kind of other empirical evidences support this hypothesis? It is not enough to present modern evidences, as it may incorrectly support a prehistoric case. A way of answering the problem is to look for pollen studies in Meso-America, and then compare them with evidences from a South American area where prehistoric southward shifts of the warm water currents and their consequent effects are measurable. The recently completed research on rainfall patterns and paleoecological studies from the marginal area of Santa Elena Peninsula can be used in conjunction with pollen studies done by Paul Sears and coworkers.[3] (Pollen profiles in the Basin of Mexico were studied.) The essential questions Sears sought to answer were: Is there evidence of climatic change within the Basin of Mexico during the time of known human occupancy? If so, can the change be correlated with known cultural horizons? While the project provided affirmative answers to both questions, the study cited in this chapter further supports the soundness of the conclusions drawn from the Mexican work.

The main conclusions that could be drawn from the preceding observations are the following. Due to cold stadial conditions and southward shift of the climatic bands, the high-pressure cells from the southwestern United States were moved to new locations over Meso-America, significantly lowering rainfall. While the data presented here is from coastal stations only, it is known that rainfall generally varies in Meso-America from about 300 mm in southeastern Pueblo to over 5000 mm in northeastern Chiapas. It should be quite possible (in another research project) to verify the hypothesis of decreasing rainfall with cold stadials in Meso-America by also collecting an annual run of rainfall figures from several inland stations.

The rainfall figures from Ancón, on the south side of the Santa Elena Peninsula, are furnished in figure 7.3.

The data run for a period of over 41 years, and is from the archives of the Anglo Ecuadorian Oil Company. In an area extremely sensitive to southward shifts of equatorial currents, the figures are striking indeed. For example, during 1968, the area was com-

3. Paul B. Sears, "Palynology in Southern North America I: Archaeological Horizons in the Basins of Mexico," *Bulletin of the Geological Society of America* 63 (1952):241-64.

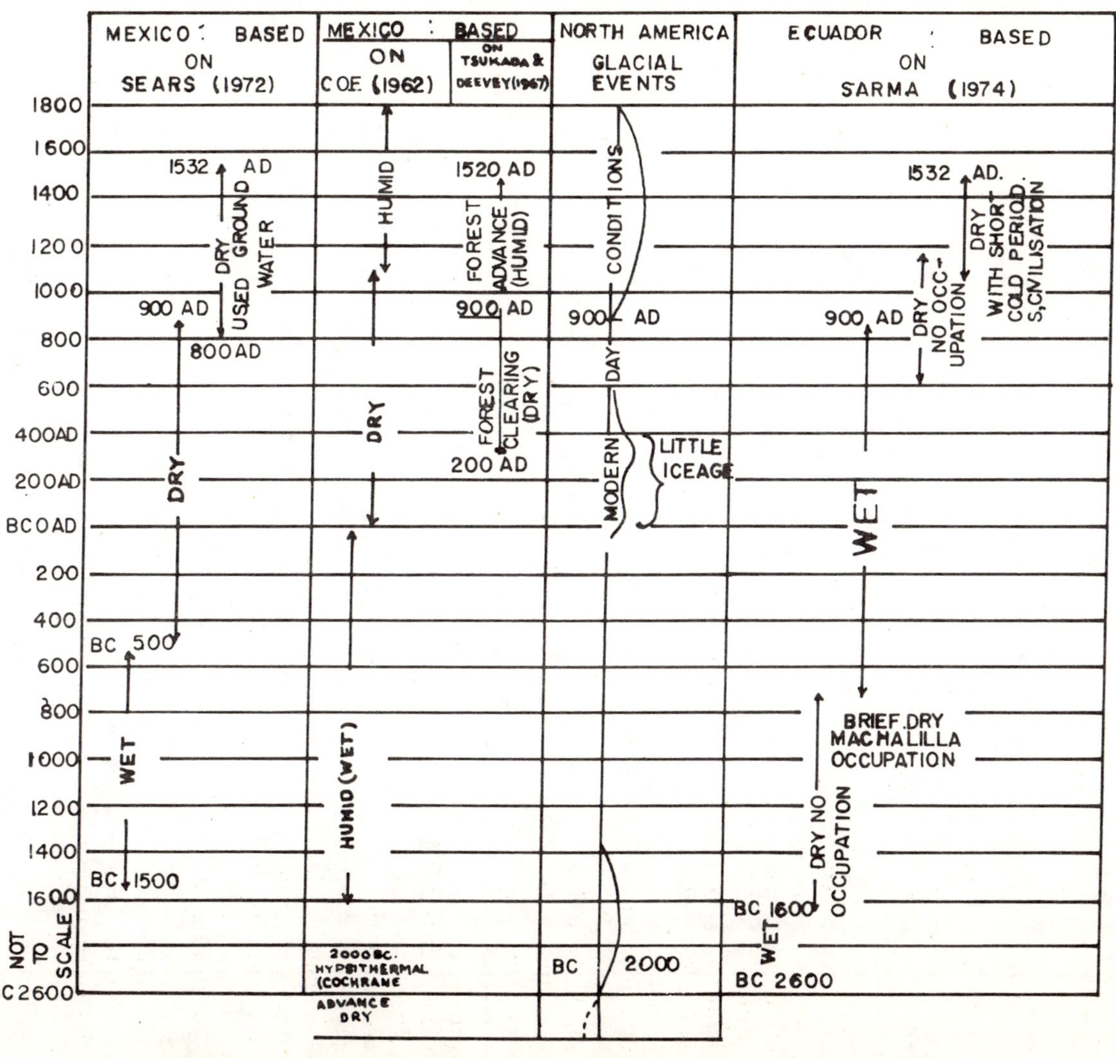

FIGURE 7.2 Climatic correlations from Central America and West Coastal America.

FIGURE 7.3 Rainfall Figures From Ancón Station, Santa Elena Peninsula, Ecuador.*
(Average yearly rainfall–212.17 mms)

YEAR	Jan	Feb	March	Apr	May	June	July	Aug	Sept	Oct	Nov	Dec	Total	±S D σ	Monthly Average
1931	1.3	1.1	3.3	1.6	0.0	0.0	0.0	0.0	0.0	0.0	0.0	0.0	7.3	.995	0.60
1932	12.9	28.7	25.9	8.9	0.0	0.0	0.0	0.0	0.0	0.0	0.0	0.0	76.4	10.221	6.36
1933	0.0	259.6	156.7	74.0	0.0	0.0	0.0	0.0	0.0	0.0	0.0	0.0	490.2	80.294	40.85
1934	63.0	45.8	59.0	8.8	2.0	0.0	0.0	0.0	0.0	7.0	0.0	0.0	185.6	23.818	15.46
1935	2.5	24.1	166.0	6.5	0.0	0.0	0.0	0.0	0.0	0.0	0.0	0.0	199.1	45.536	16.59
1936	25.0	114.0	47.2	2.7	1.2	0.0	1.0	3.1	2.5	1.0	0.0	0.0	197.7	32.427	16.47
1937	0.0	11.3	171.1	0.0	0.0	0.0	0.0	0.0	0.0	0.0	0.0	0.0	182.4	47.108	15.20
1938	0.2	69.0	71.7	1.3	1.9	4.5	1.8	1.5	0.0	3.1	0.0	2.4	157.4	25.631	13.11
1939	114.0	240.9	463.1	342.3	9.5	3.8	1.3	1.0	1.9	0.5	2.0	2.2	1.182.5	154.449	98.54
1940	85.2	31.4	7.7	23.4	0.0	7.7	0.0	0.7	9.5	0.2	0.0	0.0	165.8	23.648	13.81
1941	154.9	90.6	210.6	8.8	16.4	0.0	7.4	1.3	1.7	3.1	0.3	17.4	512.5	63.937	42.70
1942	24.8	22.4	2.0	20.3	0.0	2.1	8.3	4.5	3.4	6.2	0.8	1.3	96.1	8.704	8.00
1943	17.6	256.7	83.1	23.9	0.0	4.4	0.8	1.3	6.9	1.5	5.3	14.0	415.5	70.469	34.62
1944	10.7	47.7	109.5	8.2	0.0	3.5	4.9	1.3	2.9	0.8	0.1	1.0	190.6	99.741	15.88
1945	192.0	74.7	5.6	2.3	0.5	0.0	0.0	0.9	0.0	0.0	5.1	0.5	201.6	57.098	23.46
1946	78.8	294.6	12.6	1.8	0.0	1.5	4.9	3.1	0.0	2.9	0.0	0.0	380.2	61.111	31.68
1947	1.0	93.4	13.7	6.8	0.0	0.0	0.0	0.0	0.0	0.0	0.0	35.2	150.1	26.331	12.50
1948	12.6	2.5	0.0	0.0	0.0	0.0	0.0	0.0	0.0	0.0	0.0	0.0	15.1	3.488	1.25
1949	0.0	304.5	45.9	12.7	0.0	0.0	0.0	0.0	0.0	0.0	0.0	0.0	363.1	83.666	30.25
1950	0.5	147.5	70.1	0.0	0.0	2.3	0.0	0.0	0.0	0.0	0.0	10.6	231.6	43.126	19.30
1951	33.5	0.0	9.4	12.4	0.0	21.1	3.5	0.0	3.3	3.3	0.0	23.4	109.9	10.742	9.15
1952	32.2	9.6	14.0	35.5	1.3	0.3	23.7	0.0	15.5	0.8	0.0	0.4	133.3	12.613	11.10
1953	92.6	0.8	233.6	60.3	1.2	0.0	1.1	1.3	0.8	0.4	1.7	0.0	393.8	66.998	32.81
1954	0.0	72.8	13.5	0.0	0.0	2.8	0.2	0.0	0.0	4.0	0.0	0.5	93.8	19.994	7.81
1955	6.7	0.9	98.1	0.0	0.0	2.3	1.3	0.7	1.8	1.6	0.3	2.4	116.1	26.716	9.67
1956	30.2	3.9	44.6	4.8	1.1	0.0	–	0.0	3.4	1.6	0.1	0.0	89.7	13.797	7.47
1957	3.4	226.7	123.8	20.7	1.3	0.0	0.0	0.0	2.0	1.5	1.5	0.6	381.5	67.726	31.79
1958	15.3	139.1	91.1	19.6	2.1	2.6	0.8	2.6	0.6	0.0	0.1	0.0	273.9	42.850	22.82
1959	5.3	13.7	35.6	73.5	0.0	0.1	1.2	2.4	0.0	0.3	0.0	2.5	134.6	21.192	11.21
1960	2.3	14.5	19.7	6.6	0.0	0.0	0.0	2.3	2.3	0.0	0.0	0.0	47.7	6.246	3.97
1961	13.7	148.6	0.0	8.9	0.0	0.0	0.0	0.0	0.0	0.0	0.0	0.0	171.2	40.729	14.26
1962	10.4	0.0	3.4	0.0	1.9	0.0	0.0	0.0	0.0	0.0	0.0	0.0	15.7	2.927	1.30
1963	4.7	2.1	4.9	0.0	0.0	0.0	0.0	1.2	0.0	0.0	0.0	0.5	13.4	1.762	1.11
1964	13.5	10.9	100.4	16.3	0.0	0.0	0.0	0.0	0.0	0.0	0.0	0.0	141.1	27.367	11.75
1965	0.0	0.0	57.7	30.0	0.0	0.0	0.0	0.0	0.0	0.0	0.0	0.0	87.7	17.292	7.30

*For figures of 1966 and later, see Figure 7.1.

pletely dry. Then in 1971, the area was inundated (566.7 mm) with rainfall, as it was again in 1972 (315.4 mm). This sequential performance points out that large-scale changes involving oceanic and atmospheric phenomena occur in a time scale greater than a year. Similar extended prehistoric changes would be causal to the existence of mangroves, as discussed in an earlier section. Also, in the regions where the so-called El Niño occurs, other workers have speculated that rain occurs in a seven-year cyclical basis. But examination of the data shows nonexistence of a cycle, and that El Niño tends to occur in pairs.

IN CONCLUSION

Throughout this brief work, emphasis was placed on reconstructing the ecology of the prehistoric times by pragmatically using as many sources of information as are available. However, microecological conditions were not touched on. In any large area, the general ecology might point out an overall picture, but locally-restricted (microecological) conditions may prevail (e.g., decaying refuse in a large area). Archaeoecologically, it is important to distinguish the differences between "macro-" and "micro-" ecological phenomena even though the techniques of data collection for either technique are fairly similar.

On the whole, we have presented only a brief introduction to the fascinating subject of paleoecology. Though it is impossible to be an expert in all aspects of the subject discussed, a degree of sensitivity to the potentialities of paleoecology is nonetheless required. More than that, the subject illustrates the invaluable nature of the interdisciplinary approach fundamental to the understanding biocultural adaptation. In anthropology, this adaptation is our primary concern.

For Further Reading

Coe, M. *Mexico: Ancient Peoples and Places,* ed. Glyn Daniel. London: Thames and Hudson, 1968. Provides a background to Meso-American prehistory.

Tsukada, M. and E. S. Deevy, Jr. "Pollen Analysis from Four Lakes in the Southern Maya Area of Guatamala and El Salvador." In *Quaternary Paleoecology,* ed. H. Wright. New Haven: Yale University Press, 1971. The paper deals with pollen profiles that point out the climatic fluctuations of the Mayan area.

Bibliography

Ladurie, Emmanuel le Roy. 1971. *Times of Feast, Times of Famine: A History of Climate Since the Year 1000.* New York: Doubleday & Co., pp. 1-426.

Appendix

SOME HINTS ON PALEOECOLOGICAL DATA COLLECTION

Vertebrate and invertebrate animals as well as plants occur as both micro- and macrofossils. For comparisons, collect as many forms as possible from the region of interest.

Foraminifera: Mostly microscopic tests of ocean-core samples. From an anthropological point of view they throw light on the Quaternary period. Ocean cores are taken by special equipment, and special methods are used for study. The coiling of the shells is of importance for *climatic* purposes.

Corals: Occur in the tropics. Excellent markers for past high levels of the ocean. Corals are very sensitive to climatic changes. When labeling use strong cloth-backed labels because corals are extremely hard on labels. Double labeling (one inside, one outside) the polyethylene bags (used for storing) is a must.

Mollusks (gastropods and pelecypods): Archaeologically, derive these from screening processes if these specimens are obtained. Geologically, derive them from different strata. Undisturbed sediments should be carefully sampled. As far as possible complete, original shells are preferable, or shells that can be freed relatively easily from the matrix. Before collecting materials, learn the contemporary shell forms from the region. Thus, if different forms are unearthed, they can be meaningfully interpreted. Shells embedded in clay, shale, etc. can be released by a disaggregating agent, such as hydrogen peroxide, used with gentle agitation. Boiling is helpful. Hydrofluoric acid is also useful in releasing shells. Shells do not sustain long-term soaking. Caution is needed. In soft matrix, use brushes and forceps if the shells are strong enough. If a limestone matrix is where shells are silicified, acetic acid does an excellent job in releasing them. Thoroughly dry all wet shells before handling.

Arthropods: In archaeological contexts, these usually exist as exoskeleton bits of insects. Zinc chloride in increasing concentrations allows the chitinous exoskeletal parts to float and be skimmed off. Then they are washed and dried before analysis. Zinc chloride can be replaced with sodium chloride, when unavailable. Arthropods in geological contexts occur in bedding planes and in concretions. They are removed along with adjacent matrix, chipped away with a vibrating tool or similar equipment. Molds and casts are easier to remove with sharpened needles, gradually removing away the surrounding matters. Depending on the nature of the specimens, dilute hydrochloric, acetic or formic acids may be used.

Fishes: Can be obtained through archaeological screening processes. Fish specimens mostly occur as vertebrae and scales although other parts occur. The factors involved in the preservation of fish skeleton as a fossil are complex and poorly understood. However, these fossils give clues to ecology, subsistence patterns, and food supply. Where specimens occur in matrixes, use of dilute acetic acid is helpful. Formic acid may be substituted if acetic acid does not work.

Plants: Both macroscopic and microscopic forms occur. For microscopic materials, see palynology entry below. For macrofossils, flotation techniques (using zinc chloride) are useful. Archaeologically, seeds and fruits give the best source of information. Plant megafossils have been found in nearly all nonmarine sedimentary rocks. Generally, Cenozoic beds contain fossil leaves as megafossils. Peel transfers are usually useful in removing the materials for study. The technique needs a prepared surface (usually acid-treated). A thin film of plastic in liquid form or celluloid film dissolved in amyl acetate is then applied to the surface. After letting the film dry, peel it off. The peeling bears an impression of the macrofossil.

Palynology: Palynology has demonstrated remarkable expansion. Originally, it was largely restricted to study of Pleistocene deposits. Now, all kinds of depositional environments (brackish, lacustrine, alluvium, etc.) are being studied. The excavator can collect samples from 5 cm or smaller intervals by using a sterile test tube and pushing it into the section, making sure that there is no contamination. Lake bottoms should be expertly sampled with some type of corer. Peat deposits are one of the best sources of pollen collections. However, poor techniques of field collection will lead to later difficulties. Data should include an exact map (with such political geographic details as township, etc.), photographs/sketches, brief geological details including stratigraphy, associated fossils, and archaeological details. Leave laboratory processing of samples for an expert. A collection of pollen from present flora is necessary during beginning work at a new area. At other times, such collection is useful for identification purposes. A tenth of a gram may be adequate for pollen-rich deposits; in other cases 10 grams or more may be advisable.

Animal Megafossils: Bones and teeth of animals are obtained during archaeological screening processes. During excavation, due to age and medium in which they are embedded bones become brittle and need strengthening (with celluloid dissolved in amyl acetate solution, applied repeatedly till the material is strong enough for lifting). Teeth do not pose any preservative problems because of their hardy nature. No bone (however small or seemingly insignificant) should be discarded until a laboratory examination is completed. From the region where materials are being collected, a type collection must be established to facilitate identification. Access to similar collections from any museum is an invaluable asset.

Glossary

Artifactual Materials—An anthropological concept or idea, denoting that some material has been altered by human behavioral activity. A stone tool is an example because it shows a systematic method of manufacture.

Autecology—Deals with the study of the ecology of a single organism. This idea ordinarily refers to contemporary organisms. When prefixed by "paleo-," the term refers to the study of past life.

Descriptive Ecology—An approach in ecology, which seeks to understand complex ecological phenomena by breaking them down into simple interactions. The construction of models to simulate the conditions helps us to understand the phenomena better. This aspect is called *experimental ecology.*

Environmental Determinism—An idea dealing with environmental conditions determining cultural development. However, it does not take into consideration the uniqueness of cultural aspects.

Experimental Ecology—See Descriptive Ecology.

Fauna—All animal populations in general. In this book the term refers to animal populations of the time period under consideration, such as "Villafranchian fauna" (see below).

Flora—Refers to plants and the plant life of a period.

Functional Variability—A concept used in archaeology denoting that stone tools reflect different functional aspects: for example, tools used for butchering; tools used for digging; etc.

Geographical Determinism—A concept close to environmental determinism. Mostly used by geographers, emphasizing the deterministic aspects of the geography of the area.

Half-Life—A concept associated with radiometric dating. Here the unstable nature of an element's isotope (say, carbon-14) is used to determine its rate of decay. For example, the original amount of carbon-14 in a now-living organism will be reduced in half 5730 years from now; 5730 years is the half-life of carbon-14.

Master Chart—A chart developed by dendrochronologists for an area. They utilize growth rings and the like from such long-lived plants as ponderosa pine or bristlecone pine, develop long overlapping sequences. These charts are used (when appropriate) to date archaeological sites.

Nonartifactual Materials—An archaeological concept, referring to such materials found at a site as animal bones, transported boulders, etc. though not artifacts (tools, etc). They are nonetheless associated with past human activities.

Paleoautecology—See Autecology.

Paleosol—In cultural contexts, this term refers to the alteration of soil that takes place due to human occupation. It gives clues to the home bases of early man.

Paleosynecology—See Synecology.

Parameters—Refers to things outside an ecosystem or any system for that matter, that would affect it.

Pleistocene Overkill—Refers to the idea that man, by his hunting wiped out major mammals in several parts of the world. The concept has been widely criticized.

Processual—Refers to the way in which something goes on, such as "evolutionary process." When we think of process in hominid evolution, it is important not to think of "origin." For instance, no typical missing link between modern man and apes will be found.

Radiometry—A technique that uses the unstable nature of certain atoms for chronological purposes.

Seriation—A concept utilizing the succession of styles, say, pottery designs, for relatively ordering materials.

Settlement Patterns—Essentially a technique of analysis and observation practiced by anthropologists. Its use casts a light on how prehistoric populations occupied or abandoned an area, why and where they lived, and so on.

Stratification—A concept used by geologists and anthropologists that rests on the idea that the oldest deposits are at the bottom and the latest are at top of a given layer of earth. It gives relative sequences of an area or site.

Synecology—Study of ecological relationships of contemporary communities. When prefixed by "paleo-," the term refers to the study of past situations.

Species Equilibrium—A recent concept in ecology that points out that species are at equilibrium in general, but when disturbed, grow at an exponential rate. At equilibrium the number of animals becoming extinct will be equal to the number that would arrive in the "island" area. An island here refers to a region with demarcated boundaries; such as national parks with distinct boundaries; island ecosystems; etc.

Typological or types—An idea dealing with how the attributes or features, of an object cluster together. An example would be Stone axe. The shape, size, and manner in which a stone axe was chipped. Often this is a construct of the archaeologist. Although a useful concept, it becomes meaningless when applied too rigidly.

Variables—Different variables are present in any ecosystem. For example, the amount of animals present to the area, how man hunts them, what the animals themselves feed on, etc., are different variables.

Index

INDEX
Continued